The Hungry Dragon

This book explores China's quest for energy sources, raw materials and natural resources around the world, with a specific emphasis on oil. China's presence in Africa, Asia and Latin America is reshaping the world with regards to economics, politics and national security. This volume offers a comprehensive examination of China's energy security strategy.

The first two chapters delve into Chinese relations with energy markets and the world, and the global geopolitics of China's resource quest. This introductory section is complemented by three in-depth country case studies: Angola, Brazil and Cambodia. The two concluding chapters cover opportunities and risks to China, and examine how strategies can be developed into tangible actions.

This book offers a comprehensive number of overlapping debates regarding the varieties of capitalisms (autocratic *vs.* democratic), the urgent need for rebalancing as the world undergoes global financial crises and contestations to traditional powers, and the issues surrounding natural resource extraction in the context of global governance, neoliberalism and poverty traps.

Key Features:

- Offers an in-depth analysis on the geopolitics of China's resource quest.
- Assists students and scholars in understanding the Chinese model of autocratic capitalism and China's novel ways of securing resources across three continents.
- Explains China's energy security strategy and its implications on US national security.
- Explores the links between international relations and the geopolitics of scarcity.

Sigfrido Burgos Cáceres works at the University of South Alabama and is a consultant specializing in international development and foreign affairs. From 2007 to 2012 he was based in Rome, Italy, at the Food and Agriculture Organization of the United Nations. He lives in Mobile with his wife.

Sophal Ear is an Assistant Professor of National Security Affairs at the US Naval Postgraduate School, Monterey, California. He is author of *Aid Dependence in Cambodia: How Foreign Assistance Undermine Democracy* (Columbia Univesity Press). A graduate of Princeton and Berkeley, he moved to the USA from France as a Cambodian refugee at the age of 10.

The Hungry Dragon

How China's resource quest is reshaping the world

Sigfrido Burgos Cáceres and Sophal Ear

Routledge
Taylor & Francis Group
LONDON AND NEW YORK

First edition published 2013
by Routledge
2 Park Square, Milton Park, Abingdon, Oxon OX14 4RN

Simultaneously published in the USA and Canada
by Routledge
711 Third Avenue, New York, NY 10017

Routledge is an imprint of the Taylor & Francis Group, an informa business

British Library Cataloguing in Publication Data
A catalogue record for this book is available from the British Library

Library of Congress Cataloging in Publication Data
Burgos Caceres, Sigfrido.
The hungry dragon : how China's resource quest is reshaping the world / Sigfrido Burgos Cáceres and Sophal Ear. – 1st ed.
p. cm.
Includes bibliographical references and index.
1. Energy consumption–China. 2. Energy security–China. 3. Energy development–China. 4. Energy policy–China. 5. Natural resources–China. 6. China–Foreign relations. 7. China–Foreign economic relations. I. Ear, Sophal. II. Title.
HD9502.C62B87 2012
;333.790951–dc23
2012035251

ISBN: 978-1-85743-686-0 (hbk)
ISBN: 978-0-203-07253-0 (ebk)

Typeset in Times New Roman
by Taylor & Francis Books

Europa Commissioning Editor: Cathy Hartley

For Amy–SBC

For Chamnan, Steven and Caitlyn–SE

Contents

Preface

The book you have in your hands examines the People's Republic of China's endeavours to secure energy sources, raw materials and natural resources around the world. This examination is accomplished by looking closely at Chinese involvement in three countries: Angola, Brazil and Cambodia. The focus is mainly on gas and oil, but also on commodities, minerals, rubber and timber, among other production inputs. The guiding assumption for the chapters that follow is that, in order to sustain its economic growth and political legitimacy, China is forced to 'go abroad' in search of energy sources, raw materials and natural resources. In doing so, China is quietly reshaping the world.

In essence, China and the USA remain pivotal to energy and natural resource demand growth in the near and long terms, yet China lacks sufficient indigenous energy supplies. Consequently, to secure long-term economic growth, it must be strategic about locating where the inputs are going to originate from and tactical in figuring out how to get them to the mainland. Also, current national demand for resources has outstripped domestic supply, resulting in increased import dependence. For example, oil product demand growth will remain concentrated on gasoline, diesel and jet fuel in virtually all regions of the world as long as the rate of development of end-use technologies (i.e. private and public air, land and water transportation) drives growth forward. We believe that the current tightness in global refining capacity will ease gradually as economic recovery and capital investments pick up, but it is clear that significant capacity will be needed over the long term in the Asia-Pacific region, given that region's dynamic growth potential. Effectively, led by China, the Asia-Pacific region will consume half of global demand growth in 2012–20. These factors are major drivers of Chinese economic, energy and foreign policies.

Over the last two decades, these policies have been relatively consistent and reasonably well co-ordinated with the country's domestic priorities. However, as the world enters a competitive era marked by heightened anxiety about security of gas and oil supply, government efforts have emerged to modulate demand growth while expanding available and potential supplies, regardless of the specific governance dynamics in the location where supplies originate.

As vibrant economic, demographic, political and social trends make their way through China's mainland and its citizenry, the Chinese Government has yet to disclose high-level documents that comprehensively expound upon the country's strategic and tactical goals and the specific ways it plans to achieve economic progress and political stability through energy and resource security. While many other factors come into play in states' decision-making processes, it is evident that oil products (i.e. diesel, gasoline, jet fuel, plastics and lubricants) and basic manufacturing materials (i.e. commodities, inputs and minerals) are the essential grease that makes the economic machinery run smoothly. In view of the knowledge gap resulting from what is tangible (known) and what is hypothetical (unknown), this book aims to fill some of this gap.

As you read through the seven chapters, it will become readily evident that China behaves using mannerisms of ever-increasing assertiveness, boldness and confidence in the international arena. Beijing no longer fears belittlement by traditional hegemonic powers. Its actions in resource-rich foreign lands as well as in international forums provide irrefutable evidence that China is a country to be accounted for in international relations and respected for the size of its economy and population. The world has yet to witness the true economic power of China as it develops its domestic market. As this book is written, a great number of multinational corporations report that their most promising and biggest markets are in Chinese cities. Is this surprising? Not really. With 30% of 1,500 million people experiencing stable employment, higher incomes, social mobility and a thirst for Western lifestyles, these Chinese subpopulations are driving growth internationally just as the USA and Europe once did from the 1950s to the 1990s. Consequently, it is ever more indispensable that the global population grasp the thought process behind China's strategy and how it plans to operationalize state strategies into actions. As such, there should be continued attempts to try to envisage how Chinese energy policy could develop with respect to the goals and visions of China and its leaders, including the adaptations it makes to be perceived as a peaceful state, a good neighbour and collaborator in regional and foreign affairs.

The projected trend for China's oil consumption is upward. While some believe that China's dependence on foreign oil from 'rogue states' and unstable regions will decrease its energy security, others believe that the economic incentives will be far too strong for states and regions to play Russian roulette with China's gas and oil needs. However, as geostrategic competitions for energy sources become more bitter and frequent, there is a perceived risk that traditional hegemonic powers could block key transit points such as the Straits of Hormuz and Malacca to limit China's access to gas and oil. In response, China is rapidly building up naval forces to project power all the way into the Indian Ocean in an effort to secure gas, oil and mineral interests. With this and other issues in view, this book examines a number of overlapping debates regarding the varieties of capitalisms (autocratic *v.*

democratic), the urgent need for rebalancing as the global financial crisis continues to reverberate, and resource rents vacillate between riches and traps.

Without doubt, China's return towards the share of global gross domestic product (GDP) the country had prior to the Industrial Revolution in the West has raised questions about the USA's global and regional dominance. While it is possible to imagine a win-win outcome in which both China and the USA enlarge the pie, the reality is less rosy, as even US officials have conceded that much ground has been lost and will not be recovered. Beijing is now present throughout the entire Asian continent, in Africa, in Europe, and in North and South America. As the world hits peak oil and is slowly reaching a water crisis, China is gearing up its diplomatic and military stock to address these imminent threats on a more or less equal footing to the capacities held by other great powers. This build-up of capacities raises alarm within classical hegemonic states as it calls into question how far the instruments of hard and soft power will be used to accomplish other strategic goals and objectives. There are questions as to how this competition for resources will result in relation to success as states scramble for what is left up for grabs and, if this occurs, whom the actors vying for the lion's share of energy, materials and resources will be. While this book is about China's global search for energy and resources, it also looks at soft and hard power in light of China's new-found assertiveness, boldness and affluence.

We hope readers find the chapters engaging and informative. The narratives are easy to follow and carry the reader directly to the essence of the most pressing issues at hand. Part I, the introduction, sets the stage of China, energy and the rest of the world, as well as examining some geopolitical dynamics. In Part II, the country case studies, the actions and initiatives between China and Angola, China and Brazil, and China and Cambodia are examined in depth. In Part III, the conclusion links a number of dimensions together and examines how strategies become actions.

Sigfrido Burgos and Sophal Ear
Alabama and California, USA
July 2012

Acknowledgements

No book is ever written in a vacuum. We wish to dedicate this book to our families. This book would not have been possible without the assistance, patience, and forbearance of Amy Burgos for Sigfrido, and of Chamnan Lim, Steven Ear and Caitlyn Ear for Sophal.

Thanks to the editors of *Asian Survey*, the *Asian Journal of Latin American Studies*, the *Journal of Contemporary China* and *Geopolitics* for believing in our work when it was only in the form of essays, proceeding with blind reviews, and eventually turning our work into published articles.

Our editor at Routledge, Cathy Hartley, is gratefully acknowledged for making this book a reality and shepherding it through completion. Bob Looney (who referred us to Cathy) and Michael Glosny (both at the Naval Postgraduate School) were instrumental at various points throughout this process. This book would not have been possible without the anonymous reviewers who provided constructive comments and suggestions for what became the majority of the chapters.

Ruth Homrighaus, Matthew Copley, Jim Chhor and Richard Chhuon are gratefully acknowledged for their help at various stages of the writing of this book.

Disclaimer

The views expressed in this book are those of the authors and do not reflect the views of the Department of the Navy, the Department of Defense, or the United States Government.

Abbreviations

AIT	Asian Institute of Technology
APEC	Asia-Pacific Economic Cooperation
ASEAN	Association of Southeast Asian Nations
AU	African Union
BPD	barrels per day
BRIC	Brazil, Russia, India and China
CACF	China-Africa Cooperation Forum
CAS	Chinese Academy of Sciences
CCP	Chinese Communist Party
CEO	Chief Executive Officer
CFR	Council on Foreign Relations
ChinalCo	China Aluminium Corporation's
CIC	China Investment Corporation
CIFL	China International Fund Limited
CNOOC	China National Offshore Oil Corporation
ECLAC	Economic Commission for Latin America and the Caribbean
EIA	Energy and Information Administration
EITI	Extractive Industries Transparency Initiative
EU	European Union
FDI	foreign direct investment
GDC	Guinea Development Company
GDP	gross domestic product
GNP	gross national product
IEA	International Energy Agency
IMF	International Monetary Fund
IOCs	international oil companies
IPCC	Intergovernmental Panel on Climate Change
MERCOSUR	Mercado Común del Sur (Common Market of the South)
MPLA	Popular Movement for the Liberation of Angola
MRC	Mekong River Commission
NDRC	National Development and Reform Commission
NEA	National Energy Administration
NEB	National Energy Bureau

NEC	National Energy Commission
NGO	non-governmental organization
NOCs	national oil companies
OECD	Organisation for Economic Co-operation and Development
ONGC	Oil and Natural Gas Corporation
OPEC	Organization of the Petroleum Exporting Countries
PLA	People's Liberation Army
PPSEZ	Phnom Penh Special Economic Zone
R&D	research and development
SAARC	South Asian Association for Regional Cooperation
SAFE	State Administration of Foreign Exchange Investment Company (also SIC)
SARS	Severe Acute Respiratory Syndrome
SEZs	special economic zones
Sinopec	China Petroleum and Chemical Corporation
Sonangol	National State Oil Company of Angola
UN	United Nations
UNCTAD	United Nations Conference on Trade and Development
UNEP	United Nations Environment Programme
UNITA	National Union for the Total Independence of Angola
US(A)	United States (of America)
WHO	World Health Organization
WTO	World Trade Organization

Note to readers

Unless otherwise stated, or in quotations, all dollar ($) figures are in US dollars (US$).

I

Introduction

1 China, energy and the world

China's hunger for massive amounts of energy is driving policy in Beijing. The dynamic economic growth rates experienced in the last 20 years coupled with increased manufacturing levels, rising exports of low-cost goods, rapid urbanization, and higher demands for air travel and land transport increases China's appetite for crude oil, natural gas, timber and critical minerals.[1] In recognition of much-heralded resource depletion and talk about climate change, China has embarked upon replacing a portion of the coal, gas and oil it consumes with alternative forms of energy while addressing some drawbacks and their interference with extra growth in the renewable energy sector (i.e. solar photovoltaic, wind, solar thermal, wave movement, geothermal and biofuels from non-food plants).[2] In the meantime, Beijing continues to buy oil from traditional oil-producing countries (in the Middle East), and it also seeks its own sources abroad (in Africa, Central Asia and Latin America). These sources are sometimes located in far-flung locations, which are often, by common standards, not pleasant environments. China is also interested in minerals, such as copper, iron ore, gold, silver and tungsten, as well as natural resources like rubber, timber and potable water.[3]

David Zweig and Bi Jianhai note that China has been courting resource-rich states, "building goodwill by strengthening bilateral trade relations, awarding aids, forgiving national debt, and helping build roads, bridges, stadiums, and harbors". In exchange, China gets "access to key resources".[4] As China enters the African continent and solidifies its position as a major player in the oil industry, it gains leverage to block, either completely or partially, other international oil companies from becoming involved with upstream operations, such as crude oil exploration and production, by locking in generous accords, agreements, concession, contracts and deals with resource-rich states and getting a head start on the long lead times required to fund and develop oil in territories where extraction is logistically complicated or excessively expensive.[5]

Tellingly, China's global hunt for energy and resources is devoid of normative concerns such as human rights and democracy. Given that sought-after energy sources and natural resources are often found in nation-states with weak governance and oppressive leaders, Beijing has struck accords, agreements, deals and pacts with governments that have little to no respect for, or interest in, international norms of conduct. This way of conducting business places China at odds with American and European foreign affairs officials, because it undermines long-stipulated objectives in sensitive regions, such as isolating obstreperous and undisciplined governments or punishing them for violating civil and human rights, seeking nuclear proliferation, hosting religious radicals and terrorists, failing to comply with international law, and refusing to promote freedoms and liberal democratic practices.[6]

This first chapter asks: how and why is China searching for energy and resources globally? What are competitors saying and doing about it? These questions guide the discussion throughout, as well as inviting selected linkages to economic, environmental, geopolitical and strategic dimensions. We begin this discussion by placing the multiple aspects of China's global hunt for energy and resources into four selected dimensions: economic, environmental, geopolitical and geostrategic. Chapter conclusions and reflections offer a summary and analysis.

Economic dimension

Oil and its derivatives make the world move. Oil is used to fly planes, to run boats and ships, motorcycles and other motor vehicles, and to propel factory machines.[7] In the absence of alternative energy sources, oil and its derivatives will remain the substrate of choice in global production. Moreover, world-wide societal aspirations to attain Western lifestyles fuel an unprecedented competition for gas, forest products, minerals, oil and water, as rising—and rapidly growing—nations like Argentina, Brazil, Chile, India, Indonesia, Russia, Turkey and Vietnam pursue comfort, prosperity and economic security for more and more of their people.[8] Energy is the essence of modern civilization. Energy is the fundamental substrate that makes the world move about.

The first 11 years of the 21st century witnessed rapid economic growth and massive consumption of energy and resources by large, vibrant countries like China, but this qualifier also applies to Brazil, India, Indonesia, Russia and South Africa. This resulted in major increases in the global demand for gas, minerals and oil. In terms of fossil fuels, between 2004 and 2010, global supply could not keep pace with global demand, thus leading to a major increase in the price of oil.[9] As states recognized that shortages in supply were far into the future, they started drafting emergency strategies and subsequent measures (i.e. contingency and response plans) to deal with a world with less and less oil. One of these measures was securing a couple of months of gas and oil in strategic reserves, which is also argued to have played its part in keeping prices up. This measure was most likely borrowed from what the USA and Europe do in their territories to bring a temporal sense of energy and national security.

The rapidly rising cost of energy and resources may be squeezing oil-importing countries, but it is aiding oil-exporting states to fill their national coffers (this is especially true in Africa, where oil-rich countries have had a bonanza in state revenues that have slowly trickled down to populaces). Furthermore, China's yearly imports of energy and resources from developing countries are assisting these states in offsetting the rising costs of non-oil goods that rely heavily on gas and oil for their production, transformation and transport. Additionally, fiscal revenues in the oil-producing states, especially large producers such as Angola and Nigeria, increase as oil prices rise. Over the years, these countries have accumulated significant foreign exchange reserves—most of which have been used, regrettably, to the rulers' personal advantage.[10]

In terms of economic benchmarking, copper—a mineral heavily sought after by China—is closely watched by analysts and traders. It is called "Dr Copper" because it is a proven bellwether of the comprehensive soundness of world economy. Many industries use copper, so a strong demand for the metal often indicates that the overall economy is in the process of expansion. By mid-April 2011, copper prices had almost tripled after a two-year upswing, from $1.55 per pound in January 2009 to $4.49. This was largely driven by the belief that China, by far the world's largest copper user, has an insatiable hunger for this useful metal.[11] China's annual copper consumption hovers at 7–8 million tons.[12] Recent rising commodity prices can be partly explained by companies' and individuals' hoarding commodities of all types—from cooking oil and cotton to copper—betting that prices will increase. Experts agree, however, that successive rounds of interest-rate increases and moves to mitigate voracious speculation will have a negative impact on commodities and other critical markets in the short run. As a whole, China's long-term demand for commodities (agricultural and non-agricultural) remains robust because of the economy's size, sustained consumption levels and rapid growth.[13]

If copper is a catalyst to drive the Chinese economy, then oil usage is the economic "lubricant" that keeps the manufacturing and industrial sectors going at full speed. Beijing is becoming increasingly dependent on domestic and foreign oil, especially from Africa, Central Asia and the Middle East. This creates interstate competition and with it, "the threat of investment protectionism is growing". The tighter control of raw materials and natural resources by sovereign wealth funds and state-owned investors is a primary concern. Many "resource-rich countries are becoming increasingly anxious about China's unending thirst for direct and unimpeded control of natural resources", particularly gas, minerals, timber and oil. As developed and developing countries face an increase in the deployment of foreign direct investment from China, a growing tendency to examine deals more closely is taking place. Transactions that involve government-controlled entities, oil-for-infrastructure deals, and natural resources are being subjected to intense scrutiny by regulatory bodies.[14] A study by the Council on Foreign Relations (CFR) in New York, USA, notes that in the recent past at least 11 powerful economies—which together constituted 40% of all foreign direct investment in

2006—have cleared "new laws that would restrict certain types of foreign investment or expand government oversight" in these areas.[15]

On the one hand declining gas and oil reserves on the ground means bad business for resource-oriented companies, while on the other hand scarcity means record revenues and profits as price follows an upward trend with little indication of coming back down. This is why one of the major attractions of the oil business is its highly inelastic demand. That is, an increase in the price does not greatly decrease the demand for oil in the short run. This being the case, if oil companies can maintain a higher price for oil, they will not lose sales volume and so will reap higher profits. This partly explains why the Middle East maintains amicable relations with China, Europe, India and the USA. Traditional hegemonic powers fear that China will seduce the Middle East to prefer the East through a cocktail of incentives and indecent proposals. All countries in the world needing oil have no choice but to do business with whoever has the oil they need. America's oil import bill, for example, accounts for over two-fifths of the total US trade deficit. This is, in no short measure, a major incentive for the West to bring "democracy and stability" to the Arab world. The wave of socially led conflicts in North Africa in the first six months of 2011 may be an indication that momentum for change is picking up quickly as people are tired of political elites reaping all the benefits of resource wealth. Another source of the US deficit is the protectionist and mercantilist trade policies of major trading partners, especially those of China. China accounted for about a quarter of global growth between 2000 and 2009, outperforming the USA for the top spot as global trade-maker.[16] The evolving oil competition between the USA and China has come about because the Chinese Communist Party's (CCP) legitimacy and power, as well as the whole country's stability, depends largely on Beijing's capability for ensuring sustained economic growth above 8%, generating employment for thousands of young people joining the labour force, and providing rising living standards for its citizenry. It is under this promise that the officialdom in Beijing partly justifies its tight grip on the country. So it follows that while senior government officials are busy handling international affairs, they must also pay attention to evolving domestic (i.e. internal) dynamics which could very easily derail economic, military, political and social plans.

Attempts have been made to thwart China's growth. With the help of European allies, the USA has stepped up its heated pressure (through the media, official statements, interviews and multilateral conferences) on China to revaluate its currency. To give an idea of China's accelerated economic dynamism, one must look at exports. China's exports increased from $184 billion in 1998 to $1.2 trillion in 2007. As a result, China's trade surplus increased from $44 billion in 1998 to $262 billion in 2007, leading to increasing pressure on China from both the USA and the European Union (EU) to upwardly revalue its currency, the yuan. Yet, if the CCP allows the Yuan to appreciate rapidly—not only to please critics, but also to reduce inflation—exports will become more expensive and less competitive in foreign markets, causing exports to plunge and unemployment to rise. If inflation and dissatisfaction both continue to rise, moreover, mass unrest will follow. Evidently, China embraces free and open trade expansion, because senior officials know deeply that "if a country wants to be a world power, it has no choice but to do so through international organizations like the World Trade Organization (WTO)".[17] To this effect, China's "growing commitment to economic liberalization has increased foreign investments, commerce, and trade",[18] and it has strengthened its already privileged position in the international trading regime.

China's challenges, however, are not limited to the economic realm. The country also struggles with environmental, legal and social conflicts. William Nobrega notes, for instance, that from 1992 to 2005, over 21% of China's arable land was "put to non-agricultural use", evicting 20 million farmers from the fields.[19] Also, in environmental terms, China is now home to a large number of highly polluted cities. The following section examines these issues closely.

Environmental dimension

Environmental problems are the collective burden developed countries have to shoulder as a consequence of their energy addictions. These problems pose risks to the Chinese populace, but there are risks that remain difficult to measure. Consider carbon dioxide, an invisible, odourless, tasteless gas that is well known to cause global warming. Today, it does not appear significantly to hurt anyone in China (or its neighbours), so the perception is that managing this gas (and other pollutants) would not only be costly, but would have negligible economic payoff. Yet a World Bank study estimates that pollution (including air, land and water) 'costs China between 8 and 12 percent of GDP [gross domestic product] annually in medical bills, lost work due to illness, damage to fish and crops, and money spent on disaster relief'.[20] Similarly, China's National Bureau of Statistics estimates that health problems, environmental degradation and work days lost to pollution cost China anywhere from 4% to 9% of its total annual economic output.[21] Whereas US companies such as Dow Chemical and US Steel spend about 3% of their revenues on environment-related expenses, Chinese competitors like Sinopec and Bao Steel spend only about a 10th as much. Unfortunately, this polluting largesse comes at a steep price to Chinese citizens (they just do not know yet how bad it is). China is now home to 16 of the world's 20 most polluted cities, and of China's almost 100 cities with over 1 million inhabitants, 66% fail to meet World Health Organization (WHO) air quality standards.[22] In the end, it all comes full circle: "whatever China's pollution-based competitiveness and environmental cost advantages are at the individual enterprise level, they are offset by the aggregate social costs of Chinese pollution".[23]

Environmental degradation naturally raises a number of issues: that of sustainability, the inevitable arrival of peak oil (the point in time when the maximum rate of global petroleum extraction is reached), and the need for green energy as a solution. A number of the world's largest gas and oil fields—Burgan in Kuwait, Cantarell in Mexico, Ghawar in Saudi Arabia and Samotlor in Russia—are in decline or about to become so. Year after year, major oil producers are spending more economic resources to discover new gas and oil reserves, only to find less, and the "last decade in which new discoveries exceeded the rate of extraction from existing fields was the 1980s".[24] The solution to declining oil supplies will revolve around some combination of renewable, eco-friendly, domestically produced energy supplies, including bio-fuels, geothermal, solar, wave and wind energy generation, and less oil-dependent lifestyles in the East and West.

Inevitably, 2.5 out of 7 billion people on the planet "currently depend on wood and other biomass fuels that cause even more deforestation and air pollution".[25] So China's (and the world's) pollution problem is not limited to gas and oil burning, but to all sources of fuels, including those that long have been considered harmless and aligned with nature's way of working. In terms of international climate negotiations, some observers note that if China pushes too hard, some of its poorest, most populous provinces—those with low per capita emissions and relatively high levels of energy inefficiency—may insist on negotiating for themselves (i.e. independently) with intergovernmental organizations (e.g. the Intergovernmental Panel on Climate Change—IPCC) related to climate change and pollution mitigations to receive more favourable terms and more lenient treatment than Beijing will impose on them.[26]

Real cases are illustrative. For example, the success of Suntech Power Holding, China's leading manufacturer of modern silicon solar panels, has opened the eyes of many CCP officials to seeing renewable energy as a profitable industry that could be successfully developed. If it skyrockets, increasing supplies and investments in research and development will bring prices down so far that businesses and households will be able to afford, at least partially, to reduce coal burning and greenhouse gas emissions, replacing the lost energy with solar power. In sum, the bosses in Beijing know very well that the environmental problems on the mainland are mounting and are already taking measures to mitigate some of these concerns, but it seems that economic directives will take precedence over contamination of air, land and water. However, as the economy opens up to profitable opportunities in the alternative energy sector, it is China, with its low-cost labour and minimal regulations, which can make the most out of those market opportunities.

The next section will examine a number of changes and shifts in the distribution of power around the globe that are likely to cause tension between states in years to come.

Geopolitical dimension

For many observers, China's emergence as both a military and an economic rival overshadows environmental concerns, "heralding a profound shift in the distribution of global power. As China becomes more powerful, two things are likely to happen in the international stage: states will begin to see China as a growing security threat, and China will try to use its growing influence to reshape the rules and institutions of the international system to better serve its interests".[27] A number of commentators theorize that Beijing may be lying low for now, "until it is ready to challenge the United States in Asia and elsewhere, although China's push into" Africa, Central and Southeast Asia, and Latin America in search of energy sources, raw materials and natural resources seems to suggest that the long-expected global contest for world's remaining oil has already begun.[28] In recognition of its energy and market development needs, Beijing has influenced its southern neighbours in Southeast Asia to keep their eye focused on the ball of regional economic integration, and it has generally succeeded in overcoming the Association of Southeast Asian Nations' (ASEAN) recurrent historical suspicions of China's own hegemonic aspirations toward Southeast Asia. This is not unfounded given that Cambodia, for example, is pivotal to China's strategy to project greater influence and power in Southeast Asia through naval operations from the Sihanoukville port, buffer longstanding rivals in the region, and potentially tame America's hegemony (see Chapter 5).[29]

Beijing is not only heating up its game in Asia, but also in Africa (see Chapter 3) and South America (see Chapter 4). Let us briefly take a look at Africa. In 2000 Beijing's officialdom established the China-Africa Cooperation Forum (CACF) to incentivize commerce, promote trade and foster investment with African nations. Unintentionally, however, China's energy and resource needs may have deleterious impacts on some African states. In this book we refer to it as "The Curse of China's Thirst". For instance, "it is now well documented that increasing dependence upon natural resource exports powerfully increases the risk of conflict".[30] This is particularly worrisome in places where the costs of a typical civil war are $50 billion.[31] Additionally, unrestricted access by government officials to resource rents has significantly worsened governance in the African continent.[32]

Moreover, China is stepping up its reconciliation with Iran and Sudan, both of which are considered rogue states by Europe and the USA. It is also strengthening Chinese diplomatic and business ties to Hugo Chávez, the bad-mouthed President of Venezuela who enjoys lambasting America's foreign policies in Latin America. Far on the south of China's mainland, Australia has become a beneficiary of Beijing's insatiable quest for coal, gas, iron-ore and oil, and it has begun to explore the prospects of establishing a number of preferential trade agreements with China as well as ASEAN.

More importantly, world leaders are now considering how the magnitude of China's energy and resource needs influences the international oil market. As the energy demands of India and Brazil (and other peripheral emerging market economies) expand alongside those of China, global prices and supplies will be affected for everyone. China, Europe and the USA share an interest in stable oil prices, safe and protected shipping routes, and a secure international environment that fosters investments and capital flows, all of which can help sustain their economic prosperity and that of the rest of the world. In the end, all large oil consumers, be they old powers (e.g. the EU and the USA) or emerging ones (e.g. Brazil, China and India) share an interest in an open energy market without artificial restrictions on supplies. If such a market were achieved, Brazil, China and India would be less tempted to secure supply sources through costly bilateral deals.[33] However, the Organization of the Petroleum Exporting Countries' (OPEC) power days—although frequently said to be numbered—have not ended and will probably continue to be influential as long as there is oil to sell and someone wanting to buy it. Founded in 1965, OPEC continues to dictate the price of oil through restrictions on quantities.

Simply put, the world is witnessing a situation of dwindling gas and oil reserves on the ground while at the same time experiencing a higher demand for gas and oil from old and new powers. This competition between those who have always claimed exclusivity over most of the oil for themselves and those who are staking claims on oil and other inputs to justify their growth and success will, in the end, turn out to be a race to the bottom, a race to get whatever is left in the ground. While it is widely acknowledged that the future of energy lies on a combination of renewable energy sources, smart energy grids and non-fuel vehicles, the economic realities of exploiting old technologies until it is not numerically feasible to do so will dictate state actions.

The next section examines some geostrategic considerations of China's global search for energy and resources in a world experiencing pressures on many critical pressures and concerns.

Geostrategic dimension

China remains heavily dependent on international sea lanes to bring oil from Africa and the Middle East, and the country's leaders worry about its strategic vulnerabilities (more details in Chapter 6). It is no secret that the global politics of energy is shaping economic and diplomatic intercourse around the world. A US congressional report on Asia notes that the thirst for gas and oil is sending China (and India and Russia too) into an all-out search for energy sources, to the point that these states are appearing in parts of the world where they have not been seen before.[34] China's relatively unscathed position after the great world financial crash of 2008–09 gave it a clear and unambiguous opportunity to strengthen its strategic advantages as the USA and EU struggle to recover from a severe downturn (during the writing of this book the USA managed to avoid default by increasing the debt levels through difficult manoeuvres with a divided Congress). In terms of natural resource acquisitions, Beijing is currently in a position to assist financially through key tactical investments and direct monetary transfers to nation-states suffering from balance-of-payment deficits at a time when the West cannot come through as it has in the past.[35] Beijing has undertaken numerous negotiations in Central Asia, for instance, to back a multibillion pipeline to transport Caspian Sea oil to China's mainland.

China's economic diplomacy has beaten that of the USA, since China's Asian neighbours understand clearly now that China is their guaranteed "engine of growth".[36] However, this is far too simplistic. It comes down also to the fact that Beijing does not tell countries what to do in terms of governance and human rights; it grants them independence and non-interference. Asian countries appreciate this approach as it is diametrically opposed to what the USA says and does in terms of exhortations, threats and lectures to fellow states, acting as if America were a wealthy, vituperative evangelist. So, also, through economic diplomacy, China has markedly improved its relations with Australia, one of Washington's most faithful allies in the Asia Pacific region. Officials in Beijing believe that China's regional integration in Asia will make it easier for the CCP "to leverage economic power in support of geostrategic ambitions that could" pose a threat to US and European interests in the future. Beijing's participation in a number of regional bodies has paid handsomely in terms of acceptance, prestige and image-building. It is now easier to perceive "that prosperity-driven cosmopolitanism is superseding the biased attitude of one-party ideologues who distrust regional organizations as tools to constrain China's power and influence".[37] Despite successful reconciliations, uncertainties related to "available supplies and increased demand from emerging countries such as Brazil, China, and India have resuscitated"[38] all net-importing oil countries' fears of an overall inadequate energy security.

Alternatively, the widespread usage of renewable energy sources would reduce both dependence on oil suppliers located in volatile regions and the potential terrorist threats that nuclear plants and liquefied gas terminals attract. To avoid an energy crisis, China must cut its needs for coal, gas and oil, and replace it for new energy-generation plants via aggressive investments in efficiency. Evidently, this is easier said than done. To this end, the impact of standard setting, fiscal incentives and generous subsidies should not be underestimated given that a number of states have taken advantage of "propellers" to move their nations on the road of energy self-sufficiency. Countries like Denmark, Germany, Norway and Spain, for example, have imposed manageable standards, offered tax deductions and provided subsidies for wind power on their utilities, thus creating a market in the 1980s for wind-turbine manufacturers in Europe and the USA. Commentators insist that "rich countries should invest heavily in research and development in all energy sectors, promote conservation, develop inexpensive, feasible, and viable forms of renewable energy, and test the possibilities of significantly reducing greenhouse gases through large-scale geologic carbon sequestration".[39] A geostrategic imperative for developed and emerging market economies is the recognition that intentionally conserving natural gas and crude oil disempowers financial gains of autocratic regimes and hostile forces.

In the end, the geostrategic considerations for China and the world sit heavily on gas and oil dependence, given that these are the only outside motivators that would make China get out of its comfort zone. Beijing knows that it is the dirty manufacturing spot of the world, but in doing so it has amassed wealth for its people and acquired power for its leaders. As noted, China has enough money now in foreign reserves that it can go around "shopping" for land, forests, mines and strategic assets all over the world. At the same time, it has a powerful tool in its hands to bring all sorts of (usually poor) countries on board with its long-term plans. For many leaders, China brings the opportunities that a once benevolent USA presented to them, but this time much more is at stake. They ask: do I ensure China's energy security while getting rich by doing so, or do I ensure the energy security of my country given the limited amounts available?

Conclusions

Economically, China's growth has been impressive. Yet the legitimacy and power of Beijing and its top officials depends heavily on China's ability to ensure sustained economic growth above 8%, generate employment for thousands of young people joining the labour force and provide rising living standards for its citizenry. The need to progress economically through the continued usage of coal, gas, oil and essential minerals is what is driving China on an all-out search for energy sources, raw materials and natural resources wherever these are located.

Environmentally, the problems are daunting and serious but a number of implementable solutions are available and the desire is already in place to move China from dirty to clean, from dependent to independent, from an ecological pariah to an environmentally friendly state. While contamination and pollution are costing China about a 10th of its GDP, the calculus emerging is that it is a price worth paying for the experience of being acknowledged, powerful, bold and assertive in international affairs. The Chinese citizenry may very well accommodate an authoritarian regime, but when environmental degradation, contamination, pollution and biological devastation takes its toll on animal and human populations, this accommodation may suffer a dramatic shift as people look inwardly for their health and well-being, and not to Beijing. One has to go to Beijing and Shanghai to witness the low-lying pollution as evidence of this.

Geopolitically, the reasoning is straightforward: there is less and less of the key items that old and new powers want, and so a competition arises between powers to grab the largest share of what remains, be it gas, oil, minerals, timber or potable water. America and Europe have long-established strongholds in Africa, Asia and the Pacific, and Latin America, much like colonialism divided the world into private exploitable properties. However, what is evident today is that China and other emerging market economies are redefining the rules of the game, and are playing it with the same equipment and tools as US and European players did in the past as well as in the present. Beijing knows that the fresh scent of money with no strings attached is a powerful incentive to a large number of states, whether these are rogue or not.

Notwithstanding the occasionally media-covered diplomatic scuffle, private and public Chinese entities are leveraging their comparative advantage and competitive edge to play more cleverly the competitive game of securing energy, materials and resources all over the world. So far, this competition has been peaceful, but in the background the old powers are struggling to find ways to thwart China's drivers of growth and prosperity.

Geostrategically, it is clear that Beijing is not blind to the fact that it is under the watchful eyes of the rest of the world. It acknowledges that it has problems and that there are internal social pressures building, yet it has so far balanced all these issues through increased employment, rising living standards, better salaries, household stability, infrastructure developments, more access to information and foreign products, and openness to matters that in the past remained the exclusive purview of powerful elites. CCP officials are acutely aware of the benefits of building relations based on substance and reciprocity with allies and neighbours, and to come through all actions as a peaceful, non-hegemonic state that wishes growth, prosperity and well-being for all. This has been successfully done in Southeast Asia and the Pacific states, and the same smooth camaraderie has been deployed to the rest of Asia, a number of African countries, Russia and throughout Latin America. In the next chapters, you will read about China's inroads into Australia, Canada, Europe and the USA. This signals that the portfolio of options is wide as well as diverse, because it includes investments in clean and safe energy, and approaches that for many years were heralded as unfeasible and unviable. Today, in a world that is day by day increasing in complexity and interconnectedness, Beijing recognizes that it cannot adopt the all too familiar role of follower, but that it must become the true leader the world expects it to be.

Reflections

Nowadays, new waves of energy and resource usages are evolving in China. Over time, China will continue to build up a powerful combination of cheap skilled labour, flexible manufacturing and industrial capacity, investment-friendly central Government, and massive domestic markets. China's future is bright, with its immediate realities a little more challenging than expected. However, China's importation of gas, oil, minerals and timber from the developing world may cause a series of unintended problems. Armed conflicts are "more likely in countries that depend heavily on" energy sources, raw materials, and "natural resources for their export earnings", partly because belligerent "groups can extort the economic gains from this trade to finance subversive operations". Additionally, energy and resources fuel armed conflicts and wars, because they make secessions more likely.[40]

Realistically, the world needs to start asking more pointed questions: Will global affairs continue to be heavily influenced by Washington's free market democracy, or will it start gathering momentum from Beijing's authoritarian state capitalism? Beijing has gone to great lengths to keep democratic waves from penetrating Chinese territories. It has shut down websites, thwarted communications between citizens, banned political art, harangued against Nobel Prize nominations, and hit protesters with an iron fist. At least in the near term, the Chinese government has succeeded in stifling protest. The wave of democratization that has swept through North Africa has rippled in authoritarian regimes around the world, and China is not immune to such incendiary popular movements. Yet, "authoritarianism in China is of a far higher quality" and with more breadth "than in the Middle East". This does not mean that Beijing is protected from violent revolutions; revolution is still very much a possibility in China. However, if revolution indeed becomes a reality, it will not "come from the disenchanted poor, but from an upwardly mobile middle class fed up with anachronistic government that does little but keep" the productive social classes from achieving their true potential. This is where the revolution will start, igniting other groups to join forces and bring democracy to a land ripe with opportunity. When one-party rule becomes stagnant or fails to keep the masses pacified, change will come one way or another. One problem for China is that political unrest will surely produce economic stagnation—or something worse.[41]

Commentators are more sanguine concerning resource competition—an optimism, whether rightly or wrongly, drawn from Beijing's discourse of peaceful rise and development. There is much money to be made from conflict and war, however, and the race to secure those remaining sources of energy, raw materials and natural resources is the *ideal ignition* for the American and European military-industrial complex to lobby its respective governments to find any excuse to leverage friction, sore spots and tension that could precipitate yet another war of giants. The question is, will the world be better off with a busy China working in peace or a pressured China working in war? The deployment of troops by old powers in Egypt, Iraq, Libya and Central Asia (Afghanistan and Pakistan) can be partly explained by this internecine geopolitical contest of setting strongholds in existing and potential gas- and oil-rich fields.

Clearly, far from becoming a Western-style society, China will remain a highly distinctive state, not only culturally but also politically. Its rallying cry is: it can be done without democracy! With its continental size, massive population and restless inventiveness, China is a civilization-state, the attitudes, characteristics, guidance and values of which long predate its existence as a nation-state. For CCP officials the present is welcoming a new era in international relations that is defined by the *geopolitics of scarcity*—of oil and potable water. Greater competition for critical inputs, both commercial and political, has begun to set one country against another in pursuit of finite energy, materials and resources. To this end, China finds itself in a unique situation on a number of issues, especially those related to accountability, un-participative consultations and top-down approaches. China continues to struggle on how to deal with Iran and North Korea, in addition to climate change. Its rapid industrialization makes it difficult for the country to achieve and maintain energy satiation. Technological breakthroughs and lifestyle modifications could ease the situation, but not without anxiety, controversy and some resistance.

Notes

1 Sigfrido Burgos Cáceres and Sophal Ear, "The Geopolitics of China's Resources Quest", *Geopolitics*, Vol. 17, No. 1 (2012), pp. 47–79.
2 Judith A. Cherni and Joanna Kentish, "Renewable Energy Policy and Electricity Market Reforms in China", *Energy Policy*, Vol. 35, No. 7 (June 2007), pp. 3616–29.
3 Sigfrido Burgos Cáceres and Sophal Ear, "China's Natural Resource Appetite in Brazil", *Asian Journal of Latin American Studies*, Vol. 24, No. 2 (2011), pp. 71–92.
4 David Zweig and Bi Jianhai, "China's Global Hunt for Energy", *Foreign Affairs*, Vol. 84, No. 5 (September/October 2005), pp. 25–38, p. 26, ersaf.com/pdf/china-s-global-hunt-for-energy.pdf (accessed 13 September 2012).
5 Sigfrido Burgos Cáceres and Sophal Ear, "China's Oil Hunger in Angola: History and Perspective", *Journal of Contemporary China*, Vol. 21, No. 74 (2012), pp. 351–67.
6 Burgos Cáceres and Ear, "The Geopolitics of China's Resources Quest"; in Burgos and Ear, "China's Oil Hunger in Angola".
7 "Oil Markets Explained", *BBC News*, 18 October 2007.
8 Thomas L. Friedman, *Hot, Flat, and Crowded: Why We Need a Green Revolution–And How it Can Renew America* (New York: Farrar, Straus and Giroux, 2008).
9 Christine Hauser, "Rising Gas and Food Prices Push U.S. Inflation Higher", *The New York Times*, 13 May 2011.
10 Nathan Jensen and Leonard Wantchekon, "Resource Wealth and Political Regimes in Africa", *Comparative Political Studies*, Vol. 37, No. 7 (September 2004), pp. 816–41; Burgos Cáceres and Ear, "China's Oil Hunger in Angola".
11 Carolyn Cui and Tatyana Shumsky, "Warning Signs for Copper Market", *The Wall Street Journal*, 11 April 2011.
12 Ivan Szpakowski, a Credit Suisse analyst, estimated Chinese copper usage in 2010 at 9.5 million tons (as cited in Carolyn Cui and Tatyana Shumsky, "Warning Signs for Copper Market", *The Wall Street Journal*, 11 April 2011).
13 In particular, demand for copper remains strong owing to solid growth in industrial production, electricity usage and fixed-asset investments. The same could be said of other minerals critical to manufacturing industries.
14 "Special Report: Beyond Doha", *The Economist*, Vol. 389, No. 8601 (2008), pp. 30–33.
15 Francis E. Warnock, *Doubts About Capital Controls* (New York: Council on Foreign Relations Press, 2011), p. 3.
16 Glenn R. Hubbard and Peter Navarro, *Seeds of Destruction: Why the Path to Economic Ruin Runs through Washington and How to Reclaim American Prosperity* (New Jersey: Pearson Education, 2011).
17 Robert E. Scott, "China Can Wait: WTO Accession Deal Must Include Enforceable Labor Rights, Real Commercial Benefits", Briefing Paper (Washington, DC: Economic Policy Institute, May 1999), p. 2, available at: www.policyarchive.org/handle/10207/bitstreams/8176.pdf. China asserts that it will abide by the WTO, but not that it will be coerced into environmental and labour standards.
18 G. John Ikenberry, "The Rise of China and the Future of the West. Can the Liberal System Survive?" *Foreign Affairs*, Vol. 87, No. 1 (2008), p. 32.
19 William Nobrega, "Why India Will Beat China", *Business Week*, 22 July 2008.
20 The World Bank, *Clear Skies, Blue Water: China's Environment in the New Century* (Washington, DC: The World Bank, 1997), p. 23.

21 National Bureau of Statistics of China, available at: www.stats.gov.cn/english/ (accessed 11 April 2011).

22 Friedman, *Hot, Flat, and Crowded.*

23 Peter Navarro, University of California, Irvine, 3. *Benchmarking the Advantages Foreign Nations Provide their Manufacturers*, Benchmarking Foreign Advantages, p. 135, available at: www.americanmanufacturing.org/wordpress/wp-content/upload s/2009/06/chapter3navarro.pdf (accessed 11 July 2012).

24 In the past 25 years, only two major gas and oil fields have been discovered: the Kashagan field in Kazakhstan's sector of the Caspian Sea and the Tupi field some 150 miles off the coast of Rio de Janeiro, Brazil. For more information, see: Michael Klare, "End of the Petroleum Age?" *Foreign Policy in Focus*, 26 June 2008, www.fpif.org/articles/end_of_the_petroleum_age.

25 Sebastian Mallaby, *NGOs: Fighting Poverty, Hurting the Poor*, 1 September 2004, www.foreignpolicy.com/articles/2004/09/01/ngos_fighting_poverty_hurting_the_poor?pa ge=0,3, (accessed 11 June 2012).

26 Doaa Abdel Motaal, "Negotiating With Only One China on Climate Change: On Counting Ourselves Lucky", *Yale Journal of International Affairs*, Vol. 6, No. 1 (Winter 2011), pp. 74–84.

27 G. John Ikenberry, "The Rise of China and the Future of the West. Can the Liberal System Survive?" *Foreign Affairs*, Vol. 87, No. 1 (2008), pp. 57–60.

28 Quotation from Bernard D. Cole, *The Great Wall at Sea: China's Navy Enters the 21st Century* (Annapolis, MD: Naval Institute Press, 2001), p. 127; Paul Godwin, *China's Defense Modernization: Aspirations and Capabilities* (Alexandria, VA: CNA Corporation, April 2001); Larry Rohter, "China Widens Economic Role in Latin America", *Washington Post*, 20 November 2004; Korby Leggett, "China Flexes Economic Muscle Throughout Burgeoning Africa", *Wall Street Journal*, 29 March 2005.

29 Sigfrido Burgos Cáceres and Sophal Ear, "China's Strategic Interests in Cambodia: Influence and Resources", *Asian Survey*, Vol. 50, No. 3 (2010), pp. 615–39.

30 Paul Collier and Anke Hoeffler, "Greed and Grievance in Civil War", *Oxford Economic Papers*, Vol. 56, No. 4 (2004), pp. 563–95, p. 580, available at: oep. oxfordjournals.org/content/56/4/563.short.

31 Paul Collier and Anke Hoeffler, "Violent Conflict", *Global Problems: Global Solutions*, edited by B. Lomberg (Cambridge: Cambridge University Press, 2004).

32 The new Extractive Industries Transparency Initiative (EITI) is designed to improve governance in countries with large natural resource rents by establishing a set of codes, norms and standards as to how natural resource export revenues should be accounted for, both by companies making the payments and by governments receiving them.

33 Aaditya Mattoo and Arvind Subramanian, "From Doha to the Next Bretton Woods: A New Multilateral Trade Agenda", *Foreign Affairs*, Vol. 88, No. 1 (2009), pp. 15–26.

34 Kerry Dumbaugh, "China–U.S. Relations: Current Issues and Implications for U.S. Policy", CRS Report for Congress No. RL33877 (Washington, DC: Congressional Research Service, March 2008).

35 Roger C. Altman, "The Great Crash, 2008: A Geopolitical Setback for the West", *Foreign Affairs*, Vol. 88, No. 1 (January/February 2009), pp. 2–14.

36 Jeffrey D. Sachs, "Welcome to the Asian Century", *Fortune*, Vol. 149, No. 1 (2004), pp. 53–54.

37 Hugh de Santis, "The Dragon and the Tigers: China and Asian Regionalism", *World Policy Journal* (Summer 2005), pp. 23–36, p. 29.

38 Aaditya Mattoo and Arvind Subramanian, "From Doha to the Next Bretton Woods: A New Multilateral Trade Agenda", *Foreign Affairs*, Vol. 88, No.1 (2009), pp. 15–26, p. 22.

39 See Friedmann and Homer-Dixon, "Out of the Energy Box", p. 81.
40 Paul Collier, "The Market for Civil War", *Foreign Policy* (May/June 2003), pp. 38–45, p. 41.
41 Andrew Bast, "The Beginning of History", *Newsweek*, 18 April 2011, p. 46.

2 The geopolitics of China's global resources quest

It is no secret that China's global resources quest is changing geopolitics around the world.[1] A little closer to the mainland, its regional presence and conspicuous desire to secure energy, materials and resources has heightened the security discourse among its neighbours, especially among Southeast Asian nations.[2] One part of the explanation for this insatiable appetite for resources is China's lack of sufficient indigenous energy supplies to sustain economic growth over the long term and that current national demand for resources has outstripped domestic supply, resulting in increased import dependence, both of which can be accounted for as major drivers of its foreign and energy policy. The other part is explained by China's energy-supported economic growth: for instance, throughout the 1980s it was oil self-sufficient, but in 1993 China became a net importer of oil and jumped from seventh largest oil importer in 2000 to fourth in 2010.[3] During this process of transforming its economy—while at the same increasing its foreign gas and oil dependency—it has generated negative externalities: China is now the world's largest emitter of carbon dioxide in absolute terms,[4] though not on a per capita basis. As mentioned in the preceding chapter, the environmental concerns lingering over Chinese officials are a repeated source of friction among its own people and with foreign parties (e.g. Organisation for Economic Co-operation and Development (OECD) members) wishing to raise awareness on the delicate and potentially disastrous consequences of this matter.

In recognition of its ever-rising resources needs, Beijing assiduously seeks to secure its foreign gas and oil supplies by proactively pursuing strategic bilateral relations with key energy producers and resource-rich countries. These countries are sometimes located in far-flung places and even in exotic geographical destinations, but this is no deterrent for the Chinese in their all-out search for energy, materials and resources. This approach, while not at all different from the one being pursued by other emerging market economies, can turn out to be problematic in a world of finite resources and of a rising number of actors pursuing them. As noted earlier, this surely will lead to conflict over the remaining resources on the ground. It is in this context that the Jintao administration is attempting to ensure a secure, stable and long-term gas and oil supply to Chinese territories. An unhindered supply of energy inputs is an "essential component of China's national economic security, and to succeed, it must diversify[5] both its gas and oil sources and the ways in which gas and oil crosses its borders".[6] This global resource quest has made China an active player and generous lender in Africa, Central Asia, Latin America and Southeast Asia. As an active player in a game largely dominated by old powers makes China a suspicious actor, one that could displace the preference allocation schemes of major oil suppliers.

For more than a decade, China's vibrant economy grew at almost 10% per annum, based on cheap labour and successful penetration of consumption-oriented Western markets.[7] It is this export-led approach that helped Beijing amass trillions of dollars in reserve that was used later to give out loans to the USA and others. The growth is shown in the data: gross domestic product (GDP) growth was 10.1% in 2004, 9.9% in 2005, 11.1% in 2006, 11.4% in 2007, 9.0% in 2008, 8.7% in 2009 and an estimated 9.5% in 2010.[8] For 2011, American investment banks trimmed China's economic growth forecasts to 9.4% from 10% previously, citing a recent run of surprisingly weak industrial and manufacturing data, high oil prices and supply constraints. Although these growth rates are impressive, there is a caveat. The "double-digit annual GDP gains in the past were relatively easy because the economy grew from a low base. These double-digit gains will be harder to achieve now that China is the world's second largest economy, about one-third the size of the United States".[9] To this, one must add that inflation is not dropping as fast as China had hoped. "China's annual inflation will peak at 5.6 % in June 2011",[10] with average annual inflation forecast to hit 4.7% by December 2011. This could put a dent in macroeconomic data, but not as much as pessimists would want to. For instance, in 2012 China's economic growth forecast is 9.2%, down from the initial 9.5% that was given at the beginning. While keeping these figures in mind, one must not forget that the principal aim of China's quest for global resources is to lock up the energy sources, natural elements (rare earths) and raw materials to sustain economic growth, stimulate job creation at home, and reduce its vulnerability to fuel shortages or price shocks.

In line with supporting evidence of China's increasing international presence presented herewith, Kean Fan Lim delivers an elegant and nuanced perspective on the drivers and rationale of Chinese economic expansion overseas. Lim makes detailed connections to broader changes by taking place in global capitalism, especially the failure of classical hegemonic actors, that is, old powers (i.e. the USA and European Union—EU) to maintain public confidence in their currencies after the fallout of interconnected financial systems. The USA continued, as of August 2011, to struggle with joblessness, low growth rates, insolvency and poor capital investments. Lim narrows two developments that are intricately related to China's broader aim to attain and secure domestic economic security. The first are the multiple policies to extend the global reach of the Chinese yuan to sustain the slow evolution of a variegated global system of capitalism. The second are the preponderant and assertive roles of Chinese state-owned enterprises and the China Investment Corporation, an influential sovereign wealth fund, with specific mandates of accessing new markets world-wide to buy essential inputs and to sell Chinese outputs. These phenomena, Lim argues, are linked to Beijing's financial and monetary calculations "given its vast holdings of dollar reserves (by the March 2012, China's foreign exchange reserves reached $3.3 trillion, almost three times that of Japan) and its macroeconomic constraints through the maintenance of a fixed-rate regime of foreign exchange", which needs to be tactically harmonized to a dynamic political economy (i.e. autocratic yet market liberal) that is ensconced within the contemporary finance-driven global economy.[11] When we start to account for the geopolitical and socio-economic implications of China's sights on faraway lands, foreign affairs and its implementation of a "go abroad" policy, it becomes evidently clear that Beijing embraces a multipronged approach to economic growth, financial stability and national security.

In pursuing these aims of seeking resources globally to spur domestic employment, better the lives of its citizenry and sustain vibrant economic growth, the Chinese "government enjoys considerable support. Overall, it is backed by the pride of a nation that is finally", after centuries of decline and humiliation, returning to global pre-eminence.[12] The Chinese are holding this accomplishment tightly, as they no longer have to be belittled and marginalized by old powers. To this end, scholars have raised the theme of "humiliation" as an important tool of social and political mobilization in China. This is, we may add, a very powerful tool to call the Chinese citizenry into accepting conditions that would, otherwise, be unthinkable in normal times. An engaging piece by William Callahan notes that "national humiliation is a common and recurring theme in Chinese public culture",[13] as well as in domestic and international politics; one that is invoked in a diverse set of circumstances according to specific needs. The liberalization of markets around the globe and a denationalization of industry have been accompanied by this notion of "nationalization of shame", as it is also called. In China, for example, national humiliation seems to be a purely domestic discourse, but its salient notion of the "rightful and long-deserved position" of China on world affairs continues to penetrate and inform Chinese foreign policy in both elite and popular discourse. The transformation of humiliation and shame into vindication and salvation has been an integral part of the construction of Chinese nationalism that is now catapulted up and down throughout social strata by the Chinese Communist Party (CCP) as a source of pride, enlightenment and victory, and also as a conveniently expedient justification for national, regional and international actions and initiatives.[14]

Paradoxically, China's emerging status as a dynamically evolving nation-state (or civilization-state) lays out the *new* growth formula of transformative capitalism[15] while staying largely autocratic.[16] This growth formula is now termed the "China Model" and countries like Cambodia and Vietnam are implementing some of its most successful features. The notion of a China

Model can be best described by its most commonly identified components: "one-party rule, an open approach to free markets, and a preponderant role for state-owned enterprises in domestic and foreign economic affairs", among others.[17] Experts and scholars express astonishment at how, in the last two decades, China's polity managed so cleverly to move from a centrally planned economic system to a capitalist free-market society "largely without having any sort of grand master plan".[18] Other, more illuminative assessments explore the transformative rise of contemporary China. For example, "through the lens of critical geopolitics",[19] a Western scholar argues that a "contradictory amalgam of Western-style nationalism and a traditional totalistic conception of world order that is reactive to, and dependent on, current world politics" have helped shape China's rise to a position of assertive and influential world power.[20] This rationale seems to debunk the conventional world political views of emergence through a natural process of a guided, manoeuvred, sequential, linear succession. Indeed, any reading of Chinese history over the millennia would recognize the "return" of China and not simply its temporal "rise". It is important to highlight this "return of China" concept given that a number of critical, conservative observers in the USA still believe that the Chinese can be stopped in their tracks.

Negative pessimists of China's rise abound. These assert that "China's growth model is unsustainable and the country faces a sudden slowdown unless it undergoes urgent economic and political reforms".[21] Clearly, this reflects Western thinking toward democracy, liberal freedoms and accountability. Not that these are bad or negative values to embrace; they just simply are not Chinese values at this moment. Critics continue to cite "ever-rising social tensions, choking pollution, a lack of functional public services, and an over-reliance on exports and investment (particularly in energy and real estate) as threats to the country's economic future".[22] However, the Chinese use every opportunity to make a very important and valid point. Many senior CCP "leaders speak openly about the superiority of socialism with Chinese characteristics, as so clearly evidenced by the country's apparent resilience in the global financial crisis",[23] and diametrically opposed to the near-death drowning encountered by America and Europe.

The existence or non-existence of a compass to guide a functional capitalistic approach underplays some of the most significant drivers of production and consumption. Let us look at the facts: China hosts 1.35 billion people; the net income of the economically active population is rising; it is the world's biggest energy user;[24] the world's fastest growing market for automobiles; the world's second largest importer of oil—closely following the USA—and it is the largest importer of aluminium, coal, copper, iron ore, nickel and steel.[25] All of these are highlighted as sources of pride and victory. To further underscore these facts, in August 2010, "China overtook Japan as the world's second-largest economy"[26] and "many in the international community already accept the inevitability of China overtaking the U.S. as the world's largest economy".[27] Red-flagging China's capitalistic success, mostly by stereotyped Western critics (in a demeaning manner, as if China's progress were a random rather than a strategic event), fails to incorporate the fact that its economy benefits from massive domestic consumption and purchasing capacity that is poised to look inwards as incomes continue to rise and as infrastructure continues to develop (combined with the quick education of its population).

In spite of its newly found economic success and long-sought international recognition, the CCP is constantly being challenged domestically by issues such as increased pollution in cities (a recurrent theme in this book), corruption allegations, rising inequality, political impunity, restricted freedoms and environmental damages, to name a few. Moreover, pessimists point out that "under China's current institutional and political arrangements, meritocracy is a prerequisite for good governance but, meritocracy has been eroded by a political culture"[28] of accommodation, complacency, sycophancy and cynicism.

Despite a number of challenges, the power-hold of China's political elite "is likely to remain secure so long as it can continue to develop China's economy while creating jobs".[29] This grasp on its citizenry has attracted comments by economists and political scientists who posit that China has to grow at a minimum of 7%–8% to avoid destabilizing unemployment, deter popular discontent and, most importantly, to maintain the stronghold of a single-party government system.[30] Based on past performance and current forecasts, it seems that Beijing will be able to ride this wave without much difficulty. In the end, Chinese economic as well as energy concerns arise from four factors: 1) a growing population; 2) insufficient indigenous resources and raw materials to sustain growth over the long term; 3) policy limitations; and 4) ideological contestation.

Overall, the reformulation of China's rise as a world power and as a preponderant geopolitical actor, outside of the conventional paradigms of realism and Chinese particularity, provide an alternative view to scholars and practitioners wishing more broadly to understand the contextual scenarios and procedural avenues confronted by nation-states manoeuvring through a protracted transitional phase from bipolar to multipolar order. The themes covered in this chapter fit quite well into this camp. Also, this chapter could be seen and understood as a critique of the way Western analysts understand China's actions and initiatives. In doing so, it takes the position of viewing China sympathetically for its courage in taking control of its destiny and the ways it tries to accomplish this repositioning while remaining tactfully aware of shifting international milieu, as well as for venturing into difficult locations where foreign resource conglomerates will not go, thus increasing global supplies of oil, gas, coal and other critical resources that could end up setting prices at affordable levels for a little longer. This chapter also views China critically, for not assigning more importance to environmental and ecological degradation issues, lack of regard for good governance in partner countries, insufficient enmeshing with, and leverage of, international institutions to tackle common global threats, and its itinerant silence on criticisms levelled by allies, foes, neutral parties and neighbours in relation to its involvement in rising regional tensions, emerging armed conflicts and pressures on ideological reforms.

The following section attempts to explicate China's geopolitics in its search for global resources and the approaches it implements to achieve its goals. The deployment of economic, cultural, geographic and historic, as well as social science notions, with reference to international politics in order to analyse China's activities abroad are, herein, applied to gain a much deeper and richer understanding of Chinese relations to the world in general, and to targeted resource-rich locations, in particular. This is initiated by laying out the overarching driver of Beijing's energy and foreign policy, which is that China knows, at the moment, no boundaries in its search for energy sources, raw materials and natural resources. This is immediately followed by a brief but conscientious analysis of a number of critical implications of resources usages, and identification and explanation of the most evident and visible approaches, tools and mechanisms applied in selected or targeted regions, and the potential outcomes of these actions. Then, without trying to reach any major conclusion on a rapidly evolving subject, we offer summarizing thoughts, reflections, ideas and opinions on threats and opportunities in light of the information herewith provided. Lastly, all of the above is brought into focus to reflect on what China's resource acquisition policy and energy diplomacy mean for contemporary geopolitics and political geography. We also offer an answer to the following question: How do the rise of China and its foreign resource policy impact scholarly understandings of contemporary geopolitics? This is an important question to answer as it will elucidate the rich linkages that long have existed between critical resources, development, growth and national preponderance.

Now, let us get started with this examination by looking first at the drivers, incentives, motivations and rationales behind Chinese actions, initiatives and resulting policies at national, regional and international levels.

Drivers, incentives, motivations and rationales

It is widely accepted that economic growth is intimately linked to the production quantities of goods and services, as well as the internal and external consumption of the goods and services produced. Since it is usually measured as the rate of change in GDP, countries such as China, which necessitate a rate of growth above a threshold to thrive (i.e. around 8%) are under severe pressure to sustain these gains at any cost.[31] To be sure, there are many multidimensional limitations to economic growth, but the most critical one is resource availability. Resources encompass everything from energy sources (coal, natural gas and oil), to raw materials (minerals, timber, rubber, steel, rare earths and other essential production inputs). Naturally, Beijing is acutely aware of the vulnerability of its budding economy if it should experience a shortage of resources—most importantly of energy—because manufacturing, modern living, industrial processes and transport depend on it (more of this in Chapter 6). Simply put, a shortage of energy resources could represent economic suicide,[32] and this fact alone is what drives China's actions overseas; thus, emerging as the overarching driver of Beijing's economic, energy and foreign policy—and a source of helpless irritation to the West.

China's vulnerability to energy shortages poses an imminent threat to its economic well-being. Psychology—and also Chinese history—suggests that human populations are, for the most part, risk averse (or threat phobic), with some societies more so than others. Therefore, it seems reasonable that the CCP has responded to the threat of energy vulnerability with a three-pronged energy-securing strategy that is informed by geopolitical and geostrategic considerations (a deeper and more sequential analysis of strategies and actions can be found in Chapter 6). First, enhance domestic energy supply through new exploration and production in its territories. Second, secure foreign energy supply through bilateral and multilateral distribution and transportation arrangements. Third, diversify sources of energy supply from as many regions of the world as possible.[33] From an historical standpoint, it is worth noting that in East Asia, Japan was the first to develop a "multiple sourcing strategy" for very similar reasons. China is not at all unique in this regard and Western analysts could greatly benefit from looking at their maps for these cases before dwelling on Chinese policies and moves.

The third prong of the energy securing strategy places increasing importance on locking readily available and price-competitive foreign energy supplies as part of sustainable, long-term economic development prospects. To provide a numerical context to this approach, in 2009, for example, China imported nearly 4 million barrels of oil per day. Comparatively, this amount is 2.75 times lower than the USA, which imported 11 million barrels of oil daily.[34] To give more emphasis to this point, it is enough to recall that record crude oil imports in 2010 were 4.7 million barrels per day, in 2011 were 4.95 million barrels per day, and by August 2012 had dropped to 4.35 million barrels per day (this figure represents a decline of 15.7% from July 2012 and 12.5% from August 2011),[35] figures which have only added impetus to Beijing's actions to reassess and reformulate its energy securing strategy. This strategy receives significant feedback from international energy markets and foreign political forums.

In these circumstances, the strategic combination of economic, energy and foreign policies is most strongly applied in Africa, Central Asia, Latin America, the Middle East and Russia. There are three reasons for this. First, the remainder of confirmed world supplies of crude oil and natural gas lie in these regions; second, significant growth of energy imports to China are already coming from these countries and regions; and third, direct pipeline connections to transport crude oil and natural gas are operationally possible from some of these locations.[36]

These reasons are supported by information from a recent report by the US Energy and Information Administration (EIA) noting that it "projects that total world oil consumption will grow by 1.4 million barrels per day in 2011 and 1.6 million barrels per day in 2012. The EIA still expects that the market will rely on both a drawdown of inventories and production increases in both non-OPEC [Organization of the Petroleum Exporting Countries] and OPEC countries to meet projected demand growth". In terms of non-OPEC supplies, which are the ones concerning some of China's most recent interests, the "EIA projects that non-OPEC crude oil and liquid fuels production will increase by 540 thousand barrels per day in 2011 and by 740 thousand barrels per day in 2012".[37] What this suggests is that non-OPEC producers are contributing either half, or more than half, of all output and that China stands to benefit the most from this situation given that its energy investments lie in these locations too.

To gain objectivity in assessing China's global resources quest and comparative energy usage, it is worth noting that the current stage of China's economic development is resource and energy intensive. However, the post-industrial economies of the West, along with Japan and South Korea (plus other selected states), have been able, through innovation, operational breakthroughs, technological improvements and steep learning curves, to decouple economic growth from resource and energy consumption. On this subject, according to Guan *et al.*, who studied the driving forces behind China's recent carbon dioxide emissions surge, the growth in Chinese exports, especially of manufactured products, "between 2002 and 2005 is the primary cause of China's (noxious) emissions growth. Half of the total growth in Chinese carbon dioxide emissions"[38] in this time period can be attributed to the aggregate growth of developing countries and emerging market economies excluding, of course, that of China itself. The latest datasets reveal that "China's carbon dioxide emissions rose 10.4 percent in 2010 compared with 2009, as global emissions rose at their fastest rate for more than four decades".[39] While on the subject, this is another good point for critics, pessimists and Western analysts to follow: the apportioning of blame and responsibility for emissions into the environment should be more evenly distributed among developed, transitioning and developing countries.

One of the persistent questions in energy debates is the following: If newspapers and television say that the world has reached peak oil capability, then where is new oil coming from and how will this be achieved? Oxford economist Paul Collier notes that two major changes are taking place in the global geopolitics of oil. The first is that "democracy is spreading to oil economies". The second is that, with better exploration technologies, oil is being discovered in many low-income countries (i.e. non-OPEC). As the world can attest, for example, the spread of democracy to the oil economies was, until recently, the "overarching U.S. agenda in the Middle East". Conversely, the rapid discovery of oil in nascent democracies (and autocratic states) is a side-effect of the attempt to free rich countries' dependence on crude oil from the volatile Middle Eastern countries. To illustrate this case, "new oil field discoveries have been made in a range of low-income countries", such as Cambodia, East Timor, Gambia, São Tomé and Príncipe and Senegal, to name a few.[40] More recently, oil has been discovered in high- and medium-income countries too. ExxonMobil announced "three discoveries at its exploration project 250 miles southwest of New Orleans [USA] with the combined potential for more than 700 million barrels of recoverable oil and gas equivalent".[41] Also, India's Oil and Natural Gas Corporation (ONGC) announced two new onshore (in India) oil and gas discoveries in its exploration schemes.

In some of these countries or territories where oil is being discovered, many of the top international oil companies (IOCs) are unable and unwilling to operate because of lower than expected deposits, operational bottlenecks, technical difficulties, high risks, lack of interest, or legal obstacles. Acknowledging this point is important because it counters US President George W. Bush's specific line of argument that China's national oil companies were crowding out oil production in new locations and aggravating global oil-securing potential by keeping 'fresh' oil off the world market. In this book we argue the opposite: China, by developing oil operations in places where such IOCs dare not go brings much more oil to the market. While it is true that a good percentage of the oil pumped out will not end up in the marketplace, by satisfying a large portion of China's demand for foreign oil it contributes, indirectly, to global oil security. Moreover, the oil that does go on trading operations in oil markets, which is a substantial amount, tends to put a cap on oil barrel prices which, through downstream effects, can and does lower diesel and gasoline prices at the pump. If this is so, the West, suffering from an oil addiction that it is finding really hard to expunge, should be more receptive of Chinese overseas oil developments. In a temporal calculus, one could say that China is rebalancing some of the operational deficiencies and structural impediments that, for a long time, old powers dare not address in order to avoid friction and inconvenience. Certainly, in an environment where oil is scarce and rivalries can be easily set aside, there is now evidence of alliances and partnerships developing between IOCs and Chinese (private and public) gas and oil firms. As noted earlier, China plays the game.

As implied above, more advanced detection and exploration technologies are applied and used by energy companies, resulting in newer natural gas and oil deposits being discovered every year, deeper and further afield. These discoveries have economic, diplomatic and political implications for guest and host countries. Logically, as narrow interests converge, so does reconciliation. The resulting phenomenon of cordial relations between petroleum-rich countries and state-controlled oil companies is called "petrocapitalism". This term has been used liberally in books, columns and essays by Suzana Sawyer, Erin Freeland, Tony Hodges, Elmar Altvater, Thomas Friedman and William F. Mitchell, among others. Its origins can be traced back to the early attempts of political economists with thematic specializations to understand the multi-dimensional links between petroleum, capitalism and the state. For the purpose of better understanding the ensuing discussion in this chapter, petrocapitalism is said to operate successfully through an intertwined oil complex "that is territorially constituted through oil concessions". This oil complex, once functional, generates substantial resource rents and has "strong centralizing effects at the levels of the state".[42] Both American and European scholars are not entirely convinced by the loose boundaries of petrocapitalism and prefer to understand it more as "petroimperialism". In some academic circles the use of the word imperialism is errant. This raises two questions. First, which empire? Second, when the empire falls, as all do, will the phenomenon cease to exist? Here we do not explore these questions, but it is sufficient to note that the close energy-specific relations and deals between countries were in the near past called petrocapitalism and, nowadays, the more politically correct term "energy diplomacy" is increasingly used to describe these variegated relationships. The importance of noting this lies on the fact that it is not idealistic that the USA could propose (in the next round of the US–China or US–Asia dialogue) that energy-thirsty engines of global growth and economic prosperity join together to discuss how they could cooperate in reducing their foreign gas and oil dependence. It would be uncouth to have a misleading term as the main source of conversation. Energy diplomacy is a softer and more politically appropriate term that incorporates the tactical consideration that China, Europe and the USA are oil energy price takers—not price setters—in a world where there are only two kinds of actors: price setters and price takers.

Etymology aside, the art of energy diplomacy has been quickly learned by energy analysts and strategists in Beijing. Admittedly, China's international energy relations have been studied in detail by Western academics since the early 1980s,[43] but it is not until recently that these energy-specific foreign relations have gathered enough geopolitical substance to merit serious attention. In relation to the symbiotic nature of petropolitics,[44] Western commentators worry that strict adherence to a retrograde policy of trying to maintain control over petroleum or its derivatives in oil-producing countries purposefully delays and thwarts new innovations in high-end technologies, biotechnologies and renewable energy systems, which "could potentially shift the world into a conservation economy".[45] Other commentators believe that the future lies in sustainable energy alternatives:[46] whether or not these enjoy the support of old and new powers, they will become a necessity, brought down by force.

What troubles Chinese government officials mostly is that industrialized nations and other emerging market economies are vulnerable to random, unexpected events affecting global energy markets (e.g. natural disasters, financial meltdown, wars, etc.). Given that China is a net oil importer—one source reports that the Chinese burned 8.96 million barrels of oil per day in June 2012[47]—and therefore bound to neighbourly camaraderie, it is expected by logic and reason that it should bring its national strategic interests closer to those of the oil-dependent Western hemisphere, especially if it were to need bilateral or multilateral assistance during a brief energy shortfall. However, reason and logic do not always coincide with stated positions on foreign affairs or agreements on policy coordination. This tactically convenient alignment of interests is, indeed, only an option among others. China could also, for instance, easily decide to forge closer ties with major OPEC members of the eastern hemisphere, while maintaining or increasing its already strong relations with non-OPEC states. However, as European history of mid-seventeenth-century and American events of the 1990s tell us, the dynamic evolution of geopolitics through time gives way to dramatic changes in stances that once held utmost validity. For example, in 1990, "when the US mobilized an international coalition to drive out Iraqi troops from Kuwait", China abstained from participating in this military intervention.[48] Later, in 2007, at the United Nations (UN) Security Council, China vetoed a Western-backed draft resolution that "called on Myanmar to release all political prisoners".[49] More broadly, the ease of nuclear deterrence after the Cold War between the Soviet Union and the USA is another example of shifting political stances and ideological positions. Thus, as the examples illustrate, alignments of tactical nature are indeed tricky, but circumstances do alter cases.

Meanwhile, the USA, the EU and Japan are lobbying an ambitious, resource-hungry dragon in the form of China with the notion that securing energy supplies and raw materials for all requires a cooperative and coordinated energy and foreign policy.[50] To this end, the IEA, an energy forum composed of 28 advanced economies—to which China does not belong—could be a starting point to engage China. The IEA is "committed to taking joint measures to meet oil supply emergencies".[51] Also, the IEA member countries "have agreed to share energy information, to coordinate their energy policies and to cooperate in the development of rational energy programs that ensure energy security, encourage economic growth, and protect the environment" and natural ecosystems. By petitioning China to join, the coping mechanisms to deal with global oil supply disruptions could be improved. While this is a wildly optimistic idea, it is unlikely that China will acquiesce. It seems that only after Beijing is granted a more predominant role in international affairs, future crises and disruptive episodes might elicit more supportive and sympathetic stances from China to concerns and issues that hold sway in the West.[52]

In addition to recurrent lobbying by old and new powers, Beijing is reminded by national and international demographers that China's declining birth rate is limiting the supply of youngsters entering the labour force. This phenomenon carries geographic and socioeconomic implications. A shrinking labour market is not welcome news for a rapidly expanding economy: "with fewer replacement workers available, employers can no longer threaten to fire workers" as a whole. Armed with bargaining power, Chinese industrial factory workers are no longer willing to accept bad working conditions and meagre wages for backbreaking labour.[53] Higher wages and better conditions could rapidly translate into higher production costs and lessened competitiveness of products abroad, especially since Southeast and South Asia are picking up speed where China has left off. Also, this can be a driver for illegal immigration that can spur ethnic strife as well as heavier fiscal burdens for the state in terms of health service and utilities provision. Older workforces become wiser and more attuned to sensitive issues that creep into social circles, just as Chinese leaders are manoeuvring through information technologies (i.e. Internet access) creeping into mainland households, while in the process sowing the seeds of imminent change in the minds of younger generations that are going to carry the burden of today's acts.

In practice, and in consideration of the above, China's resource geopolitics are grounded on an acute need to match domestic resources demands with national and international supplies. In doing so, it protects the very basic drivers of economic growth—that is, consumption and production. Here, the understated caveat is that China is not alone in this high-speed race. It competes with its peers, namely Brazil, Russia, India and other in-transition economies, in addition to the usual protagonists—Europe, Japan and the USA (i.e. established powers). Concerning this topic, experts say that the ensuing geopolitical rivalries represent "one of the main obstacles standing in China's way for regional agreements on resource" extraction.[54] Not everyone believes this to be true, however.

In the midst of uncertainty, what seems to have a ring of truth is that a novel, forward-looking energy diplomacy, along with great power diplomacy and good neighbour diplomacy, constitutes Beijing's multilayered "new diplomatic approach".[55]

Economic and socio-political implications of resources usage

In the interest of broadening the discussion on the geopolitics of China's global resource quest, this short section has three specific aims that are relevant to this topic. First, it presents absolute costs for energy sources, relative costs of energy per kilowatt-hour for three sources, and the dynamics of resource usage by economic sectors. This takes care of the economic implications, as they relate to power generation, transportation (air, land and water) and processing (agriculture, commercial, industrial, processing, etc.). Second, it presents figures on supply/production and demand/consumption of resources that inform about volatile dynamics in the marketplace. This is important to underscore in light of Western criticism levelled against China in terms of unmitigated thirst for resources. Third, it briefly examines some socio-political issues that arise from unchecked resources usage and the potential consequences to Chinese leadership if emerging concerns are not properly addressed. As a whole, this section is intended to give readers a sense of the more peripheral, yet relevant, aspects of resources acquisitions, usages and wastages as they pertain to domestic affairs and international forums, especially of China's global search for energy sources, raw materials and natural resources in a more competitive world.

Let us begin with obvious issues. Coal is cheap and abundant in the countries that demand the most energy: China, India, Russia and the USA. However, oil is preferred over coal because coal produces more carbon dioxide per kilowatt-hour of energy than any other resource.[56] This is a well-accepted environmental fact, but where are the economic rationales that so often invade decision making? Why use coal and not natural gas as an energy source? How do comparative economic valuations result in relation to success? The truth is that not all energy sources are equally expensive: for the same amount of energy, coal in the West is 13.75 to 27.5 times cheaper than gasoline.[57] To illustrate this statement, and to make comparisons, the relative *costs of energy per kilowatt-hour* from three different sources—with their respective absolute costs—are presented here: coal costs 0.4 cents to 0.8 cents at $40–$80 per ton; natural gas costs 3.5 cents at $10 per million cubic feet; and gasoline costs 11 cents at $3.70 per gallon.[58] Simply put, solely with regards to cost, for the same amount of energy, coal is clearly cheapest; these are the statements that, based on economic rationality, so often inform the decision-making processes of mercantilists, industrialists and top politicians wishing their firms, factories and provinces to produce more for less, so that their dreams can become a reality faster. What are these dreams? Succinctly, these include economic progress, financial growth, social development, employment generation, electoral support, international recognition and stable political environments, among others. This simply captures the first point, of which resources are important to whom and for what reasons.

Now, let us delve into the less obvious. There is a need to distinguish, for example, between power generation, transportation and processing. When it comes to power generation, coal is cheaper than gas and oil, but is a more powerful pollutant. Clearly, oil is not the fuel of choice for power generation; coal is. Hence, in terms of power generation, comparisons are between coal versus gas, nuclear power and renewable energy sources. Above, there is no mention of the absolute and relative costs of nuclear, air, solar, geothermal and wind energy because China is not "going abroad" in search of these; therefore, its inclusion adds little to the breadth of this discussion. Comparison of this sort can be found elsewhere. Returning to our point, when it comes to transportation, oil is the preferred energy source, not because of cost, but due to convenience. There is no mass demand for gas-powered cars as they are difficult to fuel and tanks are burdensome to carry. Also, there is no installed infrastructure to deliver retail gas as there are gas stations to fill hydrocarbon-powered vehicles of all sizes. Again, this matters because China's middle-class incomes are rising and, nowadays, acquiring a car or a motorcycle is within their means. It must not be forgotten that China has 1.35 billion people brewing with great expectations for their lives and their country and they have mutual pride in being Chinese. Thus, with respect to energy processing, gas and oil are preferred over coal. A combined cycle gas-powered energy generator at a transformation plant or factory is cheaper, more efficient and far cleaner than a coal-based power plant. These application-specific distinctions are critical before making broad assertions about China's resource acquisitions and usages. These distinctions are important to understand the driving rationales behind decisions to back one non-renewable energy source or another, or to push investments in renewable energy alternatives.

The dynamics of supply and demand in the marketplace play an important role in countries' decisions on acquisitions and usages of national, regional, or international resources. Specifically, where supplies of oil are concerned, the rapid economic growths of China and India have absorbed most of the available oil-pumping capacity, meaning that demand has, by now, outstripped the ability of the existing oil deposits to maintain production. In this respect, this is one part of the explanation for China's insatiable resources appetite. Furthermore, China's newfound car culture—as the USA and Europe have learned—must be constantly fed[59] and, coupled with China's newly built factories and mega infrastructure projects, requires ever-increasing fossil fuels that today account for 12% of the world's energy demand and four times the global rate of oil consumption.[60] This results in higher crude oil prices, meaning that country-specific decisions have ripple effects on consumers world-wide (it should be noted here, too, that the same is said of pollution and greenhouse gases).

Oil producing countries are fond of China and India given that, for example, Saudi Arabian oil can be pumped for $5 a barrel and sold for 10 times more to rich nations like the USA, Canada, France, the United Kingdom and Germany.[61] The few points on supply and demand dynamics, detailed above, are the type of information that should inform Eastern and Western analysts, as well as political geographers and international relations scholars with regards to acquisition and usage of resources and its concomitant layered impacts on socio-political structures.

China's energy sources, however, have not changed much over the last decade. Coal has long dominated mainland energy consumption, accounting for 70% of total energy supply. This volume is massive and to reduce this dependence on coal a series of major actions and policies must take place so they take effect 20–30 years from now. Gas and crude oil are responsible for 21% of total energy supply, with the remaining 9% for nuclear and renewable energy sources. By returning to a very basic analysis of China's energy mixture and what is driving demand for coal, gas and oil, it is easier to understand that because coal is abundant and inexpensive, China utilizes coal-based power to meet about three-quarters of its energy needs, while constantly building more energy-generation plants throughout its territory.

With respect to the above consideration, one can conclude that coal is only cheapest when ignoring the carbon dioxide intensity of coal-fired power generation; also, coal mining has a high human cost in China and abroad. As if emulating weak supply, demand for coal-based electric power is exacerbated during extreme weather periods. For instance, it is estimated that the aggregate number of air conditioners consumes almost one-third of electricity demand in China during summertime. Cold winters can consume up to two-fifths of electricity output.[62] Again, this is important to recognize in analyses, given that forecasts on rising temperatures owing to climate change are between 0.5 to 2.0 degrees Celsius. If this is indeed so, the world can expect a lot of people in China (and the rest of the world) to be using air conditioners, which will further increase production of noxious atmospheric gases if the majority is concentrated on coal-based generation.

We now turn our attention to the consumption and demand of selected energy sources or natural resources for 2009 and 2010 in different locations, in order to get a better idea of how all the dynamics result, in relation to success, when compared and contrasted between countries.

Overall, in terms of consumption and demand of natural resources, when comparing China to the USA and Europe (see Table 2.1), one notes that in terms of crude oil, China consumes 2.04 times less than the USA and 2.84 times less than other OECD countries (excluding the USA). This suggests, and this is largely known and acknowledged by energy analysts, that the West is far more oil consumption-oriented than China, which should not be surprising. In terms of dry natural gas, both Europe and the USA consume on average 6.9 times more than China. Again, this should not be surprising, given American and European urbanization and its need for gas to heat homes, businesses, schools and hospitals. Also, because of the importance of USA and Europe and their proper infrastructure, in both seaports and inland, to handle vast quantities of liquefied gas, both tend to consume more of it. Furthermore, Russia uses gas distributions via its grid over Ukraine to pressure the EU. Lastly, and in line with our brief analysis of the relative costs of energy per kilowatt-hour, coal is the preferred energy source in China: up to three times more than the levels of Europe and the USA.

Chinese coal reserves (114.5 billion short tons of recoverable coal) are estimated to last more than a century, even when one accounts for expected increases in economic growth and industrial expansions. This explains why China is not seeking coal overseas, for it has an abundance. So great is China's fondness for domestic coal that, according to the EIA, it is currently

Table 2.1 Comparison of crude oil, dry natural gas, and coal consumption/demand for 2009 and 2010 by country and/or region

	*2010**	*2009*	*2009*
Country/region	*Crude oil million barrels per day*	*Dry natural gas billion cubic feet*	*Coal thousand short tons*
China	9.31	3,075	3,474,665
Europe	–	19,324	943,711
Other OECD**	26.44	–	–
USA	19.01	22,816	1,000,424
Total world	86.36	106,760	7,577,379

Notes: * Average demand by the second quarter;
**Organisation for Economic Co-operation and Development (OECD).
Source: US Energy Information Administration, www.eia.doe.gov.

building about 1 gigawatt of new coal-based energy generation, in the form of coal-fired power plants, every week.[63] Plus, China imports coal into coastal markets because its internal railway system cannot supply sufficient coal to meet demand. China's large coal-powered growth could, if unrestrained, irredeemably damage more local ecosystems, disrupt its economic drivers, and endanger the rest of the world. Is this not what the failed Copenhagen Climate meeting was all about? More poignantly, what is the international community to do with a recalcitrant India and an aloof Russia? For its part, the Chinese officialdom is highly aware that climate change represents a major threat to ecologically vulnerable areas in the highlands that are the sources of potable water, such as the Qinghai-Tibet Plateau. These perturbations may translate into less water feeding into rivers that could also decrease significantly the potential to generate hydroelectric power, which is also a cheap energy source. It is now evident that "China cannot afford to do what the West did: grow fast now and clean up its mess later".[64] In short, the problem for China's energy usage combination is its overwhelming reliance on coal, followed by its rising dependence on foreign gas and oil. In reaction, the Chinese Government has laid out plans to expand gas-fired power generation, as well as substantial investments to increase the contribution of renewable energy sources to total domestic supplies. In our opinion, this is the correct course to take given that very little time is left for experimentation and wishful thinking.

As noted in the introduction, the legitimacy and robustness of the CCP to pursue its domestic and foreign objectives depends chiefly on maintaining rapid economic growth. This means that strategies to secure energy sources, natural resources and raw materials to feed this growth are overplayed, whereas environmental regulations that get in the way of growth tend to be underplayed or plainly ignored.[65] To a degree, the disappointing declaration of the Copenhagen Climate Change summit in December 2009 is partly a result of multiple-point disagreements between China and the USA, the most significant of which are related to carbon dioxide emissions and other noxious air pollutants.[66] The world is taking note of this complacency; meanwhile time is running out and patience is dropping short. Also, the transformation from "shame and humiliation" to "pride and glory", noted earlier, could halt entirely and resolutely should a dramatic shift in climatic phenomenon take place. Anyone who has been to Beijing or Shanghai can attest to the heavy pollution that impedes viewing of what lies a few yards ahead. Moreover, this world-wide phenomenon of fast information delivery to constituencies modulates the depth and breadth of geopolitical and socioeconomic policy instruments that any country can deploy at any given time. Specifically, the rapid and massive growth of communication technologies in Asian populations increases the internal economic, political and social pressures on Chinese leaders and other authoritarian regimes. Some of these pressures are specific, such as human right violations, poor working conditions, lack of channels to voice grievances, rising pollution and ecological degradation;

others are general, like, for example, currency manipulation, international responsibilities and unambiguous stances on conventions and treaties. As increasingly networked electorates bask in this so-called "political awakening", officials in Beijing find that the room for manoeuvres in dealing with distant and near countries, as well as their local constituencies, is vastly constrained. *Realpolitik*—so cleverly practised by Otto von Bismarck and Henry Kissinger—is nowadays the most common prescription in utilitarian foreign affairs and, in line with this trend, China repeatedly reassesses aspects of its newfound leadership at a time when critics to this leadership have weakened across the globe.

In the end, the emerging concerns mentioned above, if not properly addressed, could have potentially detrimental consequences to Chinese leadership inside and outside their country. So, finally, one is forced to ask, what is driving foreign gas and oil demand in China? Not power generation, clearly. It seems very likely that petrochemicals, air, land and sea transportation, industrial processing, specialty-products manufacturing, and strategic stockpiles are feeding into calculations and ideas directed to China's combination of economic, energy and foreign policymaking.

The following section brings our attention to how China seduces resource-rich countries into acquiescing to its proposals and suggestions, and how old powers react when ententes are signed and sealed and announced bombastically by media outlets. This seduction is done through state courtship, outright contestation, tactical manoeuvring and playful tactfulness.

Courtship, contestation, manoeuvring and tactfulness

Throughout history, a combination of political and geographic factors have influenced domination and power, and fuelled many conquests and expeditions to foreign destinations. Now, as in the past, rulers and leaders alike are in an identical position to display outward actions and emit signals that either strengthen or weaken their prestige, status and wealth with respect to others.[67] This has manifested already in countries engaged in armed conflict over energy. The past speaks for itself: in 1941, when Japan invaded China, the USA responded with an oil imports embargo. To Japan, oil was a critical input. A retaliatory action from Japan was expected. On the morning of 7 December 1941, the attack on Pearl Harbor took place. Now, try to mentally fast-forward 61 years: the Iraq war (or operation Iraqi Freedom) was launched in part because Iraq is the US keystone to the Middle East oil region.[68] As you can see, it is not a recent phenomenon that countries fight over energy and natural resources.

In the recent past, a number of journalists and scholars have commented on and studied the geopolitics and economic diplomacy of China's oil quests in Africa,[69] Asia Pacific,[70] Central Asia,[71] South America,[72] Southeast Asia,[73] North America,[74] Russia[75] and elsewhere around the world.[76] Because energy and resource deals are immersed in a dense fog of secrecy, speculation and misinformation, there is lag of six to 12 months before reputable think-tanks and scholarly journals can confirm and/or release the facts and report their analyses on the range of potential repercussions arising from particular tactical policies or strategic moves. The global media, however, with its vast network of reporters and investigative journalists have better access to local events and indigenous narratives that more easily permeate the news sources from which the public gathers information and builds knowledge. Largely from newspapers, reports and articles do we learn of China's numerous reconciliations with foreign governments and ownership stakes in multinational energy corporations and natural resource enterprises.

Some countries act boldly and faster than others do when they learn of sweeping moves that threaten general and/or specific interests in a particular region. For instance, the Carter Doctrine, which stated that the USA would use military interventions if necessary to defend its national strategic interests in the Persian Gulf region, was a response to the 1979 invasion of Afghanistan by the Soviet Union. This response was intended to deter the Soviets from seeking hegemony in the Gulf. While it is unlikely that the USA would launch a similar response to China's moves under the more centrist and moderate administration sitting in Washington (to be sure the US invasion of Iraq could also be seen as a pre-emptive move to contain China's global resource quest while at the same time keeping closer tabs on Iran), it still leaves room for speculation in terms of pre-emptive actions if heightened competition for critical resources were to ensue. It also opens up questions on China's expansion into Latin America, long considered America's backyard under the Monroe Doctrine (more on this in Chapter 4).

In the following paragraphs a series of handpicked examples are presented to demonstrate how courtship, contestations, manoeuvrings and tactfulness play out in relation to the Chinese world-wide resources search. Pay attention as to how Beijing goes about closing deals swiftly.

In North America, current and former senior US officials have stated that Canada's tarry rocks and sands play a pivotal role in US energy security; however, Chinese state-owned oil companies have strategically acquired substantial stakes in Canadian energy companies and plans are already underway for pipelines from the province of Alberta to Canada's western coast for delivery shipments to Asia. Although there are limits to the extent of foreign ownership in Canadian energy firms, it is widely believed that Beijing continues to acquire minority stakes in a range of small, medium and large resource ventures.[77]

In Latin America—which has long been considered the USA's preserve and thus residing within its sphere of influence—things have changed dramatically during the last five years as a result of China's rise. Chinese energy companies have followed an aggressive investment policy in South America, with a focus on Brazil and Venezuela in particular. Here are some of the details: in Brazil, investments and loans to the order of $8 billion are financing infrastructure developments such as ports, rail, aviation and mines.[78] In Venezuela,

President Hugo Chávez confirmed that China agreed to extend $20 billion in loans, a sign that economic diplomacy is deepening ties between the two countries as Beijing seeks to secure energy supplies in key oil-producing countries.[79] This amount is on top of the $16 billion already invested in gas and oil exploration, production and transportation infrastructure. Colombia and Chile have seen an influx of Chinese investments, albeit small compared to Brazil and Venezuela, to finance investments in mineral exploration, particularly of copper, iron ore, gold and silver.[80] For a more detailed discussion on Brazil and Latin America, see Chapter 4.

In the Middle East, Chinese leaders regard a number of Middle Eastern countries as volatile, unstable and subject to strong US influence, even victims of outright manipulation. For years Beijing has been moving away from the Middle East and towards Africa. For instance, Saudi Arabia's share of total oil imports by China has been declining from 18% in 2005, to 16.4% in 2006, and to 14% in 2008.[81] Furthermore, US-led efforts to isolate Iran diplomatically or pressure it to give up its ambitions for nuclear weapons and uranium enrichment have been complicated by China's past and present commercial agreements with the Islamic republic. To give substance to this context, for instance, in October 2004 Tehran signed a $70 billion oil deal with Beijing. It is clear from the reconciliation that the evolving ties between these two countries could undermine America's overarching national security objectives.[82]

The vast African continent has emerged as the centrepiece of China's overseas energy strategy, with Angola playing an increasingly dominant role as its crude oil supplier.[83] China has increasingly dynamic economic and diplomatic relations and political activities on the continent. The leadership sitting in Beijing, aided by cunning and seasoned regional strategists, believes that the Chinese vision of alternative economic development and chequered historical experiences resonate melodiously with African counterparts, and that the longstanding history of development cooperation and amicable political linkages enables a durable and robust foundation for future bilateral partnerships—especially as it relates to land and natural resources. Beijing, understandably, does not place all its eggs in one basket. To further diversify its oil sources, Nigeria signed in May 2010 a $20 billion deal with China to build oil refineries and energy infrastructure which, arguably, continues to accentuate Beijing's preferred emphasis on resource-rich, governance-poor nations.[84] However, according to Power and Mohan, in Africa, China acts as a geo-economic and geo-political actor that is simultaneously different and similar to other interventionist industrial powers. Their rationale is "premised on a political economy approach", which weaves narrow material interests (e.g. oil, gas, minerals, raw materials) with a tailored "deconstruction of the extra-economic ways through which Chinese capitalism" becomes internationalized.[85] Another important player is Sudan, a place where "China's largest energy company is pumping crude oil through a Chinese-built pipeline to the Red Sea, where oil tankers wait to ferry it to China's" manufacturing and industrial cities.[86] However, it would be misleading to say that China is sucking dry all of Sudan's oil. According to the US Energy Information Administration, in 2008, a little over half of Sudanese oil went to China and one-quarter to Japan. Moreover, in Sudan, the actions and behaviours by Chinese actors have sometimes been a balancing and moderating force in the country, while provoking contestation, strife and violence in rather unlikely places, such as Zambia. This surprising swap is contrary to popular perceptions. Despite all the misinformation revolving around anything relating to China, there has been much interesting commentary and "media interest on the impacts" of Sino-African engagements on governance of Africa's nation-states. These engagements are "often seen to be strengthening authoritarian states, such Sudan and Zimbabwe". While there may be some truth to this, however, few remarks and annotations are seen or heard about China's engagement with "more democratic states and spaces, such as, for example, Zambia" and others.[87] The evolving debates over the so-called mixed blessings of resource rents at the hand of Chinese gas, oil and mineral imports in the context of governance, liberal-oriented notions, and poverty traps in Africa and Central and Southeast Asia tends to be overplayed in light of the desire to negatively portray Beijing's action overseas, and also because fast communication technologies permit dissemination of information to a wide number of interested stakeholders.

Western scholars note that China's embrace of the African continent is long-term, planned, strategic and flexible as situations unfold. Its tactical engagements are delivered through aid, direct investments in multiple sectors, technological transfers, bilateral trade agreements and special economic zones. After all, Beijing's experiments have "lifted millions of Chinese out of poverty" and they believe that the application of these same tools can help African countries to develop.[88] Chinese officials use their newfound success as a hook to gain sympathy (and a little admiration) from African partners. Of course, to Beijing, sending aid to Africa and Latin America opens up new markets for Chinese exports, and the threat of denial could be used to exert political and economic pressure. These intentions can be summed up by an Eskimo proverb: "Gifts make slaves just as whips make dogs". Marcus Power, however, calls for a more sympathetic and understanding view of Chinese practices. He ascertains that China is "uncomfortable and unfamiliar with the notion of development as an independent policy field". In this sense, the economic manoeuvres of China as a lender of last resort to belligerent and corrupt states has earned it the term of rogue creditor, dictatorship supporter and an increasing threat to the international aid industry. In view of this, a careful rethinking of China's geopolitical roles in development assistance is urgently warranted.[89]

Despite subtle geopolitical contestations, China is cultivating good neighbourly relations, fostering military alliances and building deepwater ports throughout Central and South Asia[90] because it recognizes that four-fifths of its oil imports pass through the 1,100 kilometre-long "Strait of Malacca—a shipping channel between the Indian and Pacific oceans—that represents the shortest sea route for China's" African and Middle Eastern energy imports. Heavy reliance on a single shipping lane has its drawbacks. This waterway is vulnerable to terrorism, piracy and naval vandalism. Also, a more pressing concern for Beijing is the dangerous threat of foreign intervention. China is desperate to "dilute its reliance" on Malacca, since a military-backed blockage of the strait is a feasible and viable intervention that the USA (and its allies) can undertake should irreconcilable differences arise between China and the USA. Unsurprisingly, "direct protection of Chinese trade routes and other interests in the Indian Ocean" is supported by the Communist Party's leadership, through the establishment of a mighty naval "fleet able to project power over long distances", with the express aim of protecting key Chinese vessels.[91]

To this effect, "part of China's new economic geo-strategy is to seek alternative routes for transporting natural gas and crude oil".[92] This is already occurring: by mid-2010, it had already signed a deal for a 20-metric ton transfer capacity pipeline for gas and crude oil from Myanmar (Burma) to southwest China. For its part, Myanmar stands to benefit greatly from East–West tensions in the region since it is geographically positioned for policing the waterway that concerns Beijing most. Myanmar is open to assume this role as it will probably help the country get closer to China and benefit from spill-over effects of its growth and progress. Already Myanmar enjoys economic and diplomatic support of Beijing, and this is closely linked with the need to cultivate long-lasting friendships with states with direct access to the Indian Ocean.

Moreover, China is also exploring the possibility of a highway, pipeline, or railway stretching from the Gwadar port in southwest Pakistan to its western province of Xingjian. However, this idea may take a secondary place after witnessing, in August 2010, the disastrous floods that can affect this area. Additionally, it is widely believed by Western analysts that these pipeline projects close to the Arabian Sea will do very little to minimize Beijing's vulnerabilities in the Strait of Hormuz (between Iran and the United Arab Emirates), through which two-fifths of China's crude oil imports transit annually.[93] Other transport alternatives under consideration include building a high-capacity railway across Southeast Asia, linking China to Singapore, which if agreed upon, could optimistically be completed by 2015. Alternatively, building an overly ambitious 'Panama-style' canal across the Isthmus of Kra—a narrow piece of land connecting the north and south borders of Thailand—which would allow tankers carrying crude oil from Africa and the Middle East to bypass the already heavily transited Strait of Malacca.

Also, China is making increasing usage of the recently opened East Siberian Pacific Export Pipeline, which conveniently links to distribution points in northeast and northwest China. Japan, the world's third largest economy, is also competing for Russian oil exports via this new pipeline. This could evolve as a point of conflict under the geopolitics of scarcity. Japanese officials are no different in their reasoning than the Chinese, as they share the same ideals for alternative energy transport pathways: avoidance of the bottle-neck of Hormuz and Malacca, and dilution of potential discord with the consumption-oriented Western hemisphere. Energy experts allege that the fundamental underlying rationale behind all of China's energy and resource deals with Angola, Iran, Nigeria, Russia, Sudan and Uzbekistan is that it allows Beijing to avoid direct confrontation with Western interests—especially, not having to see eye to eye with the USA.[94]

In 2006 Erica Downs argued that "China's international oil and gas investments were unlikely to bring Beijing the energy security" it desired.[95] Clearly, much has changed in the geopolitics of energy since that year. Chinese national oil companies are laboriously scavenging for resources, materials and opportunities in the global oil and minerals market. In Beijing, the political apparatus is deploying a monumental empowerment of diplomatic engagements to build up efforts to foster ties with oil-rich countries, secure bilateral supply contracts, promote alternative transportation routes and provide greater access to overseas investment opportunities. Evidently, the Chinese Government is increasingly synchronizing its actions, efforts and policies with those of state-owned oil companies.[96] However, diplomatic overtures and policy engagements, by their very nature, take time to bear fruit. Throughout 2011, these evolving realities will provide signals to Beijing that it has diplomatic opportunities and geostrategic leverages[97] to confront Europe and the USA that were not available five years ago.

Additionally, the Chinese have been investing in research and development at their investigation centres and universities to develop renewable energy alternatives, but they have accomplished few successes so far. One of these alternatives is "algae fuel. Algae fuel might be an alternative to fossil fuel and uses algae" (or cyanobacteria) "as its source of natural deposits. The production of biofuels from algae does not reduce atmospheric CO_2, because any CO_2 taken out of the atmosphere by the algae is returned when the biofuels are burned. Also, they potentially reduce the introduction of new CO_2 by displacing fossil hydrocarbon fuels".[98]

Under increasingly complicated scenarios, the geopolitics of China's global resources quest is further complicated by the portrayal of China as a voracious creature willing to stamp out those who dare get in its way. This raises trepidation among world leaders about the apparent contradictions between the often communicated notion of a more multilaterally minded China and one that is predestined by its sheer size and momentum to soon lead the world. Furthermore, the popular media and fast Internet transmission have helped affix the notion that global warming is viewed by many Chinese as an orchestrated conspiracy by the West to slow China's growth.[99]

In short, as this section demonstrates, Beijing has been staying disciplined and focused on its global quest for energy sources, raw materials and natural resources to sustain China's continuing high economic growth and political legitimacy. This quest is driving domestic economic, energy and foreign policies as well as modulating contemporary international politics. In complement to the abovementioned narratives, a more nuanced elucidation of China's recent "go abroad" policy is a necessary antidote to the often stereotyped Western critics of China's resource acquisition strategies and the ways it uses them.

A multipolar world order for energy

In the last two decades the world's largest economies have exhibited a preference for medium- and long-term measures to energy security that are anchored on a market-based approach—that is, the reliance on the efficient functioning of national and international energy markets (oil, gas and coal) to satisfy their needs.[100] "On the other side of the spectrum, China's measures to enhance security of energy" supplies are strategic and tactical in nature. This strategic approach combines "state-sponsored economic measures with salient political initiatives": direct government participation in domestic energy production and overseas energy investments, trade-for-aid deals, domestic and foreign development of primary and secondary infrastructure, low-interest credit products, and tailored subsidies, among other things. Countries pursuing this strategically tactical approach, such as China nowadays and Japan in the 1970s, are not overly concerned with the costs of policy implementation when these are compared with the probability and potential economic (and political) costs of resource shortfalls or energy supply disruptions.[101] In retrospect, Japan's energy decisions helped its successful economy to improve the life and well-being of millions of Japanese in urban and rural locales throughout. The Chinese and its top officials look around at successful examples and ask themselves: if they did it, why would we not also do that?

The balanced combination of market-based and strategic approaches to energy security is believed to render the best outcomes. In Asia, rapid economic growth has had positive and negative spill-over effects on abutting as well as distant countries. One effect of this accelerated growth is that Asians are increasingly desperate to secure supplies of natural resources, regardless of location, prices, or limitations. The Chinese, in particular, hop throughout Africa, Asia and Latin America, searching, negotiating and securing natural resource deals with individual countries, but so is everyone else. This is the reasoning behind the notions of resource competitions getting more and more vitriolic. Demand-side stimulation of energy markets, most obviously oil, is at first sight very good news for resource-rich countries around the world because it raises prices of their export commodities, thus swelling their coffers. However, leading economists worry that these higher prices for resource exports are likely to slow the urgency for reforms in countries with bad governance and poor economic and social policies. Additionally, the widely acknowledged "resource curse" increases the risk of conflict in countries with a history of armed conflicts, rebellions, coups and belligerent political contestation.[102] Lastly, this market stimulation raises the possibility of disruptions in an attempt by major players to tilt the balance of power, which is intimately linked to energy control, in their favour—as well as against their opponents.

Bilateral and regional cooperation between Asian countries is deeply rooted in growing regional trade, rising energy demands and security concerns. The many energy diplomacy issues that arise from these interactions seem to be best dealt with within a multi-cooperative framework that includes "all interested" parties.[103] Examples of such collaborative frameworks in Asia are the Association of Southeast Asian Nations (ASEAN),[104] the South Asian Association for Regional Cooperation (SAARC),[105] and the Asia-Pacific Economic Cooperation (APEC),[106] among other smaller arrangements. The participative nature of this collective well-being thereby lends itself to minimize potential discord in a multipolar world; however, as conflict between North and South Korea demonstrates, discord remains under the surface and ready to erupt.

It is relevant to the harmony of global commons and urgent need for rebalancing as the world undergoes global crises also to acknowledge that the economic development of newcomers, spurred by globalization, raises concerns involving issues of culture, environment, resources, societal progress and power dynamics. These concerns, which are also important to emerging economies, are now thought to necessitate a cleverly contrived scheme that combines diplomatic, economic, legal, political and social measures to tackle them at global levels. In the past, rich countries (i.e. old powers) paid little attention to the world's poor despite accruing the resources to lend a serious helping hand. It is believed that the rise of China, and the leaps of other emerging market economies, can be a political opportunity to advocate for redistribution policies that favour developing countries, especially of well-governed, resource-rich countries wishing—tactically and in an orderly manner—to penetrate export markets in a sustainable fashion.[107]

In addition to donations that have trickled down from the First World, emerging market economies like China, and to a certain extent India and Brazil, have been recently active in the international development scene as donors to the developing world, and even participate in formal consultative group meetings in China's case, as happened in Cambodia. The aid sent has, nevertheless, been comparatively little to what old powers have been giving. Another avenue to deliver "aid" to countries in need is to buy their primary commodities. Given that the price of natural resources has followed an upward trend, this has been greeted with enthusiasm by the governments of Africa's and Central Asia's resource-rich nations. From the perspective of the resource exporting states, China's "trade as aid" is very attractive as it comes with no obligations, where as development aid from the West does. However, not everyone buys into this trade as aid approach. Some commentators find this notion rather naive, as it tries to mask unequal exchanges between parties.[108] To make matters worse, it seems to be locking resource-exporting countries into a neo-colonial relationship with China.[109]

The alternative to resource-pegged aid is delivering massive increases in the economic resource transfers to Africa, Latin America and elsewhere.[110] This would be hard to justify, if at all possible, as any poverty-afflicted province or state within a donor country would express disdain. However, still pertinent is that as emerging market economies become more active players in international affairs, the weight of their responsibilities also becomes much heavier.

It would be wrong, though, to place a heavy development burden on China when it is still struggling to cope with sizeable domestic problems. It must, therefore, not be forgotten that at this point in time China's vibrant economic system is geared toward an undervalued exchange rate to promote cheap exports, rather than expanding consumer credit in order to boost domestic consumption. Although emphasis on exports is benign for market economies, it raises issues of dependence: China must export the goods it produces to keep afloat. China cannot manage to absorb its throughput domestically, in part because more than 600 million urban citizens live on an average of about $7 a day, while 750 million rural people live on an average of $2 a day (compared to the federal minimum wage in the USA of $7.25 per hour). This export-oriented strategy works smoothly when there is endless demand for the goods it makes. This demand, for the most part, comes from the USA, Europe, Australia and Japan. Economic strategists believe that over-reliance on export could soon result in "burnout". In the interest of comparison, Japan's economy depends on international trade for approximately 15% of its GDP. In China, on the other hand, that figure is 36% of GDP and rising.[111] To this end, and from a geopolitical standpoint, China is well advised to: avoid as much as possible pestering its largest clients with irksome issues; move strongly to lessen competitive actions and connotations; and continue to seek bilateral and multilateral energy agreements so that the focus is aimed elsewhere, for example at India or Brazil.

In this context, when over one-third of aggregate national output depends on international trade, the primacy of fixing production capacity on an upward trend is rational. Even at a plateau, production for internal and external markets calls for gargantuan use of energy sources, natural resources and raw materials. Here, we compare two time periods that can illustrate the dramatic changes in energy consumption efficiencies. The first time period is called resource-efficient growth: in the 20 years between 1980 and 2000, while China's GDP grew fourfold, its total energy consumption doubled. Rightly so, the second time period is called resource-inefficient growth: in the four years between 2001 and 2005 energy consumption grew twofold over GDP growth as the Chinese embarked in energy-intensive build-up of its nation-wide infrastructure.[112] Moreover, by 2003, "China's energy consumption amounted to 1,678 million tons coal equivalent (MtCE)" and was estimated to reach 2,173 MtCE by the end of 2010.[113] These values are set to continue to increase in years to come.[114]

Tellingly, from 2001 to 2010, most of the growth was resource inefficient or, as other more polite authors also using cement production as an example put it, growth has been "exceedingly yet unavoidably" resource intensive. In reality, China has embraced alternative energy sources through large state subsidies, but these only deliver a small share of overall needs. For example, cement production is energy wasteful and with so much of the country under construction, the rate of cement consumption would increase rapidly. Currently, the Chinese equivalent of America's interstate highway system is being built from the beginning, with modern China appearing to be made out of concrete. Also, buildings and apartments are mushrooming everywhere. Unsurprisingly, it is estimated that nearly half of the world's cement is produced and used in China, and cement factories, as implied before, are a major source of the country's surging demand for energy and resources.[115] To this effect, the Chinese Government is undergoing industrial assessments that aim to improve the energy and carbon intensity of its economy. For example, rather than agreeing on carbon dioxide emission targets, China is solidly committed to reductions in carbon production intensities.

Nowadays, blue-collar Chinese may very well think they sit atop the world. This is only possible because the CCP's dominance and influence is felt in all walks of life as it seeks not only to hold absolute control of the legislative and executive bodies, but also to dictate the country's moral guidance and ethical values. Indeed, the party remains the most powerful nation-wide institution in China, far ahead of the People's Liberation Army (PLA).[116] It is important to stress this point, given that tidal, unipolar mass movements rely so much on information management and ideological manipulations. As China opens up to the world, it is inevitable that the modern world will flood China. As it does, its populace will quietly start to reassess the validity of the one-party rule, antiquated restrictions and other sensitive matters (e.g. Tibet, human rights, Taiwan's annexation, civil freedoms, potable water crisis, etc.).

For its part, the USA has been cautious in advocating balanced bilateral trade, reforms to intellectual property laws, improved governance,[117] currency revaluation and upholding human rights,[118] all of which are tension-provoking issues. Part of this caution arises from the fact that China is the world's biggest foreign owner of US dollars (i.e. 70% of its total foreign currency reserves, estimated, by June 2011, at about $3.25 trillion), and also the largest holder of US government debt. These figures suggest that China holds a very large financial stake in the USA. Similarly, Beijing is heedful of unnecessary invective towards Washington, since it has a vested interest in America's national welfare.[119] In addition to these facts, Central Asia is the oil-rich region closest to China and where Afghanistan, Iran and Pakistan are geographically located. Eastern and Western foreign policies need to acknowledge the need of the Central Asian leaderships to strike a balance between American, Chinese and Russian interests in order to uphold regional order, preserve their sovereignty and avoid operationalization of multipolar conflicts in their territories.[120] The policy coordination needed is said to require convergence, fairness, inclusiveness, respect, volition and political will, as well as reciprocal courtesy and friendship of nations marked especially by mutual recognition and respect of executive, legislative and judicial acts to uphold critical national interests.

As part of the international architecture for influence, the Chinese Government desires a high level of protocol when associating with old powers. Beijing craves recognition by the international community that China is an important, powerful, dependable and consultation-worthy country.[121] However, acceptance at high protocol levels often comes with a few added responsibilities, most of which are related to adherence to international standards and norms.[122] Not many Westerners see China adhering to these and, as a result, China's desires for an uncontested position at a high level may be curtailed by machinations of its own design.

In relation to energy-pertaining foresight, "guidance concerning research directions, the planning process", and investments into future resource geopolitics, the examples can be mined from strategic reports from the Chinese Academy of Sciences (CAS)[123] that "draw a framework for eight basic and strategic systems of socio-economic development with the support of science and technology" (including biotechnology) which are steeped in ecological stewardship.[124]

To summarize this section, we chose British environmentalist Tristram Stuart's groundbreaking book *Waste*, which succinctly notes that "there is now more reason than ever to view society's interests in a global perspective. It is no longer rational for the rich to deplete natural resources regardless of where they are in the world. Doing so harms those local environments and indigenous people, and it deprives others of resources needed for survival, which is no longer morally tenable, if it ever was".[125]

Concluding remarks on the geopolitics of China's global resources quest

This section seeks to narrow down and urgently stress some relevant points on a topic of great contemporary interest. It also informs scholars, economic and foreign policy practitioners, political geographers, historians, international lawyers and diplomats on a broad range of national and international policies (and subsidiary policy instruments) that are shaped by specific goals and vision which have, and will continue to, influence the changing world map, as well as the interactional dynamics in a rapidly fluid international order.

First, China's transformation from a peripheral, isolated, agrarian society into a major economic force and a key player "in the global economy has spawned a voracious appetite for" energy sources, "raw materials", and other critical resources to sustain its economic growth at about 8–10%. To satiate this hunger, Beijing is sending its private and public companies to faraway lands and distant points of the globe in a scramble to conquer global resources before others do. This approach has been termed "assertively competitive" and traditional hegemonic powers might view these interventions as aggressive, yet little is said or done about it. While venturing into foreign lands,

Chinese companies enjoy unconditional support from their government and from a coordinated foreign policy that promotes, mainly through national oil and mining companies and sovereign wealth funds, the securing of resources in the international energy market. Without any doubt, energy security is of central importance and it has, therefore, been the driver behind China's "outward-looking" strategy.[126] Indeed, much of our contemporary lifestyle has been designed around cheap gasoline, but at current oil prices, the price consumers are willing to pay for fuel depends not only on the energy it delivers, but also on its convenience.

Chinese leaders might choose to secure and obtain the energy products and raw materials needed throughout China in the cheapest possible way so that the competitive advantages and other resources they currently possess can be devoted to improving education, health, nutrition and the overall economic well-being of their citizens. Addressing sustainable resource usage is a global public good; however, providing it is going to be somewhat problematic. This and other "global public goods are grossly undersupplied because no-one has much interest in providing them. Being good for everybody, they face the ultimate free-rider problem".[127] Notwithstanding, the actions of a few countries can exert significant impact in areas that are not well understood. For instance, in 2007 a research article was published showing that dust and soot from coal plants in China could affect the local climate of the Pacific Ocean enough to change circulation patterns of water and wind.[128] This provides evidence that the tragedy of the commons has gone global. Also, this is evidence that actions by foreign actors can have impacts in faraway lands which can put the pressure on internationally to mitigate these unintended effects. However, the question is, how is China to reconcile the grievances of others without any scientifically rigorous and indisputable evidence to support claims? Evidently, one can only manage what one measures, and China is not interesting in measuring negative externalities when nobody has been interested in doing so in the past. There is strong reason to believe that Beijing will remain mute to US and European pressures on externalities and other legalistic claims.

Second, geo-strategically speaking, China's sore spot is the open vulnerability of the Strait of Malacca, the unique waterway that allows for 80% of oil shipments to reach the mainland. This trepidation is not unfounded. An American fleet of warships sits readily at the Eastern end of the strait, at the Changi Naval Base in the southernmost point of Singapore. At the distant Western end, a renowned US-UK naval outfit operates from the tropical, footprint-shaped coral atoll of Diego Garcia, in the British Indian Ocean Territory. In response, and quite intentionally, China is pursuing military and naval capabilities to operate as far as the Indian Ocean and further into the Pacific. In fact, China's annual defence budget has grown at 14% per year.[129] The US Department of Defense, the Pentagon and State Department officials interviewed by journalists in Washington note that recently "the Navy of the People's Liberation Army is investing in nuclear-powered submarines" and its first aircraft carrier to support additional missions beyond a potential Taiwan contingency,[130] which very likely means protection of oil tankers and other vessels along the Straits of Hormuz and Malacca. While Beijing is assiduously seeking alternative energy sources and multilateral cooperation to strengthen its energy security position, it is also using its available naval forces, ballistic missiles and air force to extend its power projection to areas where the USA is currently also seeking more partnerships and stronger alliances.[131] We believe it is only a matter of time for commentators to start equating the current China-US economic and military tug-of-war with the grotesque suspicions and mistrustful features of Cold War tensions that arose between the USA and the Soviet Union. Also, a replay of the Carter Doctrine is still suspected. To be sure, the language may have changed from "strategic competitors" during the George W. Bush Administration to "strategic partners" during the Obama presidency, but the relationship's outlines remain the same.

Ideally, if the present economic recovery moves ahead, the financial windfalls accrued by the BRIC economies of Brazil, Russia, India and China will, in theory, raise the net incomes and standard of living of roughly 3.5 billion people, more than half of the world population. However, while the utopia of lifting millions out of poverty is compassionate and benign at its core, it carries socio-political, agro-ecological and geo-economic implications with lasting consequences for generations to come. One thing is hardly debatable: in view of the massive energy demand imposed by China becoming in 2009 the world's largest automobile market, both by sales (rising 45% to 13.6 million units) and by production (13.8 million units), it will not take long for China rapidly to develop a high-throughput transit system for its burgeoning middle class to move from home to work and back. Western critics could grant China a little credit for its entrepreneurship and creativity, as well as for its nascent global responsibility roles.

Conveniently, too, African leaders and politicians mightily embrace the flow of aid[132] from China that comes without Western lectures about human rights, liberal civil freedoms, environmental stewardship and good governance.[133] True, exchanges may be unequal with neo-colonial overtones but one has yet to see evidence of African states running away from China because of this. On the contrary, the African continent is open for business, and India, Portugal, Spain, Brazil and Russia, among others, have taken very good notice of this fact.

Third, the relative prices of fossil fuels (including coal) and solar-based energy may prove to be the most important factor in China's future resource decisions. For a moment, one can pretend to wear Chinese President Hu Jintao's shoes and attempt to answer the following question: With a burgeoning population that continuously experiences and suffers from famines, lack of opportunities, poor health, malnutrition, poverty, repression and widespread illiteracy, would you stymie economic growth to give in to Western pressures in order to keep the global temperature from rising a few degrees in decades to come? Only you can answer this, but we suspect Jintao's answer is an "absolutely no". Alas, consideration of answers to this question and some of the themes discussed in this chapter point to policy changes in priority areas that could give China's present administration its best opportunity in providing a positive role.

Lastly, in the multipolar world where we currently live, China's fundamental "interdependence with the rest of the world" plays a preeminent role in nurturing peaceful coexistence,[134] accepting fair arrangements that reflect the fluid distribution of economic power, and a more balanced governance of an interconnected, transnational economic system. This can only be achieved by building and strengthening alliances, increased participation in international institutions, comity between nations, and multilateral collaboration in tackling four of the world's principal global challenges—climate change, collective management of the global economy, non-proliferation of nuclear weapons and stewardship of the UN within an international legal framework.

This is entirely possible. In the last 10 years the world has witnessed a Beijing that reacts strategically and reflexively to changes in other economies and to its self-imposed macroeconomic constraints by seizing emerging opportunities for creation of new markets in new geographical areas, as well as by constantly addressing mutating politico-economic realities around the globe. If this can be accomplished in economic and mercantile spheres, then it can also be applied to common threats and hazards in social, cultural, diplomatic and political spheres. Also, and in addition to the above, this peaceful interdependence can also be achieved by the usage and leverage of existing institutions, tightening the bonds between China's values and policies, and proactively engaging critics and opponents with clear and simple justifications based on sound evidence.

Final reflections on two questions

Hence, the reader may ask: what does this mean and why does it matter? Simply put, it means that a revisionist China will go forward as decided and planned, and that the rest of the world should start getting comfortable with the idea of accommodating non-traditional emerging powers. It matters because it informs contemporary geopolitics and political geography on the multidimensional impacts of China's foreign resource acquisition policy and energy diplomacy.

How does the rise of China and its international resource policy impact our understanding of contemporary critical geopolitics and political geography? Considering that contemporary critical geopolitical scholarship continues to be engaged critically with core and peripheral questions surrounding geopolitical discourses, practice and consequences, and the history of geopolitics around the world, the impact to the discipline is on comprehending that China's multipronged approaches to domestic needs and the subsidiary policy instruments applied are informed by multiple sources and assessments that do not lose track of the guiding principles tightly held by one-party rule. It also provides a closer look at an alternative growth model—*authoritarian capitalism*—and how its success is setting a visible precedent that is now considered a direct and viable contestation to the widely embraced Western neoliberal capitalistic model so persistently hinged on democratic values, liberal freedoms and universal rights. Finally, it hints that the economic interests of developing dream-ridden nations know no flag, no currency, no language, no approach, no skin colour, no hymn and no past. China is embraced for what it is, what it stands for, and what it offers. As the West continues to struggle with China's emergence, these chapters contribute a much broader explanation that could influence paradigmatic shifts more suited to incorporating non-traditional actors in contemporary worldviews.

Acknowledgement

This chapter is adapted from a refereed article of the same title by the authors, published in *Geopolitics*, Vol. 17, No. 1 (2012), pp. 47–79.

Notes

1 This argument has been further developed by T. N. Marketos, a lecturer at the Athens, Greece branch of the Centre d'Etudes Diplomatiques et Strategiques, Paris, France, who notes that China's need for energy has become a driving factor in contemporary world politics and a precondition for sustaining China's continuing high economic growth. See: Thrassy N. Marketos, *China's Energy Geopolitics: The Shanghai Cooperation Organization and Central Asia* (London: Routledge, April 2010), pp. 168. Previously, also, by Kjeld Erik Broedsgaard and Bertel Heurlin, *China's Place in Global Geopolitics: Domestic, Regional and International Challenges* (London: Routledge, 2002), pp. 216.

2 David Shambaugh, "China Engages Asia: Reshaping the Regional Order", *International Security*, Vol. 29, No. 3 (Winter 2004/2005), pp. 64–65; "Asian Security Co-operation: Lost Horizon", *The Economist* (London), 10 June 2010.

3 David Zweig and Bi Jianhai, "China's Global Hunt for Energy", *Foreign Affairs*, Vol. 84, No. 5 (Sept.–Oct. 2005), pp. 27–28; Paul Dibb, *Toward a New Balance of Power in Asia* (New York: Oxford University Press, 1995).

4 In 2007 China emitted 21% of global carbon dioxide emissions, up from 14% in 2002 and 8% in 1981. See: Dabo Guan, Glen P. Peters, Christopher L. Weber and Klaus Hubacek, "Journey to World Top Emitter – An Analysis of the Driving Forces of China's Recent CO_2 Emissions Surge", *Geophysical Research Letters*, Vol. 36, No. 4 (2009), Cite ID # L04709.

5 Hongyi Lai, "China Builds Bridges to Fuel its Engine Room", *Financial Times*, 4 July 2010, available at www.ft.com/intl/cms/s/0/6e350266-879a-11df-9f37-00144feab dc0.html (accessed 11 July 2012).

6 H. Harry Lai, "China's Oil Diplomacy: Is it a Global Security Threat", *Third World Quarterly*, Vol. 28, No. 3 (2007), p. 521.

7 Barry Naughton, *The Chinese Economy: Transitions and Growth* (Cambridge, MA: The MIT Press, 2007), p. 11; Thomas L. Friedman, *Hot, Flat and Crowded: Why We Need a Green Revolution and How it can Renew America* (New York: Farrar, Straus and Giroux, 2008), pp. 343–66; "China", www.state.gov (accessed 27 August 2010).

8 National Bureau of Statistics of China, www.stats.gov.cn/english/ (accessed 27 August 2010); "China's GDP Grows by Seven-year Low of 9% in 2008", *Xinhua*

(Beijing), 22 January 2009; "China GDP Grows by 8.7 percent in 2009", *CNN* (Atlanta), 21 January 2010; Qiang Xiaoji, "World Bank Sees China GDP up 9.5% in 2010", *China Daily* (Beijing), 17 March 2010; "Watching China Whizz By", *The Economist* (London), 19 August 2010.

9 Andrew C. Schneider, "Seven Flaws in China's Growth Model", *The Kiplinger Letter*, 23 August 2010, available at www.kiplinger.com/businessresource/forecast/archive/seven-flaws-in-chinas-growth-model.html.

10 "Goldman Cuts China 2011, 2012 GDP Growth Forecasts", *Reuters*, 24 May 2011, available at in.reuters.com/article/2011/05/24/idINIndia-57228020110524 (accessed 11 July 2012).

11 Kean Fan Lim, "On China's Growing Geo-economic Influence and the Evolution of Variegated Capitalism", *Geoforum*, Vol. 41, No. 5 (2010), pp. 677–88, p. 677.

12 Suisheng Zhao, "China's Global Search for Energy Security: Cooperation and Competition in Asia-Pacific", *Journal of Contemporary China*, Vol. 17, No. 55 (May 2008), p. 208; Heinrich Kreft, "China's Quest for Energy", *Hoover Policy Review*, Vol. 139 (Oct.–Nov. 2006), available at www.hoover.org/publications/policy-review. Also, in the context of interstate cooperation, it is worth noting that China has security cooperation agreements with the USA, particularly after the terrorist attacks of 11 September 2001.

13 William A. Callahan, "National Insecurities: Humiliation, Salvation, and Chinese Nationalism", *Alternatives*, Vol. 29 (2004), pp. 199–218, p. 214.

14 Callahan conceptualizes national humiliation as a "transnational model that takes particular forms, but largely within the limits of nationalist discourse as it has developed in the context of anti-imperialist revolution and civil war since the nineteenth century … intertwined not just with modern China but with modernity". See: William A. Callahan, "National Insecurities: Humiliation, Salvation, and Chinese Nationalism", *Alternatives*, Vol. 29 (2004), pp. 199–218, p. 215.

15 For a comprehensive and detailed examination of who runs what in industrializing China, see: Doug Guthrie, *Dragon in a Three-piece Suit: The Emergence of Capitalism in China* (Princeton, NJ: Princeton University Press, 2001), pp. 320.

16 Stefan Halper, *The Beijing Consensus: How China's Authoritarian Model will Dominate the Twenty-first Century* (New York: Basic Books, 2010), pp. 30–33; Michael Hirsh, "We're No. 11!" *Newsweek*, 12 August 2010.

17 "The China Model: The Beijing Consensus is to Keep Quiet", *The Economist* (London), 6 May 2010.

18 Edward S. Steinfeld, *Forging Reform in China: The Fate of State-owned Industry* (New York: Cambridge University Press, 1998); G. John Ikenberry, "The Rise of China and the Future of the West: Can the Liberal System Survive?" *Foreign Affairs*, Vol. 87, No. 1 (Jan.–Feb. 2008), www.foreignaffairs.com.

19 Waleed Hazbun, *The Middle East Through the Lens of Critical Geopolitics: Globalization, Terrorism, and the Iraq War*, 2001, p. 2, available at hazbun.mwoodward.com/HAZBUN_Lens%20of%20Critical%20Geopolitics.pdf.

20 John Agnew, "Emerging China and Critical Geopolitics: Between World Politics and Chinese Particularity", *Eurasian Geography and Economics*, Vol. 51, No. 5 (2010), pp. 569–82, p. 569.

21 Jamil Anderlini, "China's Growth Model 'Unsustainable'", *Financial Times*, December 2010, available at www.ft.com/intl/cms/s/0/f38e08ce-0e84-11e0-b9f1-00144feabdc0.html#axzz20F1K01pP (accessed 11 July 2012).

22 HSBC, "Sterling Absolute Return Strategy", *Quarterly Review* 1, October to 31 December 2010, p. 3, available at www.assetmanagement.hsbc.com/uk/attachments/investors/qr_sterling.pdf.

23 Jamil Anderlini, "China's Growth Model 'Unsustainable'", *Financial Times*, December 2010, available at www.ft.com/intl/cms/s/0/f38e08ce-0e84-11e0-b9f1-00144feabdc0.html#axzz20F1K01pP.

24 "China World's Biggest Energy User", *The Times* (London), 20 July 2010; Jared Diamond, *Guns, Germs and Steel: The Fates of Human Societies* (New York: W. W. Norton, 1999).

25 Leo Lewis, "China Demands More Say in Global Iron Ore Trade", *The Times* (London), 13 August 2009; Chris Alden and Ana C. Alves, "China and Africa's Natural Resources: The Challenges and Implications for Development and Governance", Paper No. 41 (Johannesburg: South African Institute of International Affairs, September 2009), p. 5.

26 "China: Second in Line", *The Economist* (London), 16 August 2010; David Barboza, "China Passes Japan as Second-largest Economy", *The New York Times*, 15 August 2010.

27 Jamil Anderlini, "China's Growth Model 'Unsustainable'", *Financial Times*, December 2010, available at www.ft.com/intl/cms/s/0/f38e08ce-0e84-11e0-b9f1-00144feabdc0.html#axzz20F1K01pP.

28 Yu Yongding, "China Going Forward", *New Europe Online*, 2 January 2011, available at www.neurope.eu/blog/china-going-forward-0.

29 Ken Miller, "Coping with China's Financial Power", *Foreign Affairs*, Vol. 89, No. 4 (July–August 2010), p. 96.

30 Franklin Allena, Jun Qianb and Meijun Qianb, "Law, Finance, and Economic Growth in China", *Journal of Financial Economics*, Vol. 77, No. 1 (July 2005), pp. 60–64; Also, Naughton, *The Chinese Economy*.

31 Paul M. Romer, "Economic Growth", in David R. Henderson (ed.), *The Concise Encyclopedia of Economics* (Indianapolis, IN: Liberty Fund, Inc. 2007); William Easterly, *The Elusive Quest for Growth: Economists' Adventures and Misadventures in the Tropics* (Cambridge, MA: The MIT Press, 2002), Chapter 11, pp. 217–39.

32 Countries go to war—as Japan did during the Second World War—in search of mining and energy resources.

33 Farizal Mohd Razalli, "China's Foreign Energy Quest and the Changing Security Dynamics in ASEAN", Asian Conflicts Reports, Issue 11 (Alexandria, VA: Council for Asian Transnational Threat Research, June 2010), p. 1.

34 US Energy Information Administration, www.eia.doe.gov/cabs/China/Oil.html (accessed 27 August 2010).

35 "Oil Data: China August Crude Imports at 22-Month Low", *Fox Business*, 10 September 2012, available at www.foxbusiness.com/news/2012/09/10/oil-data-china-august-crude-imports-at-22-month-low/#ixzz26HqHo2Kc.

36 Michal Meidan, "Shaping China's Energy Policy", *Journal of Contemporary China*, Vol. 18, No. 61 (2009), pp. 603–4.

37 US Energy Information Administration, www.eia.gov/steo/ (accessed 2 August 2011).

38 Dabo Guan, Glen P. Peters, Christopher L. Weber, Klaus Hubacek, "Journey to World Top Emitter", *ISEE*, 2010, p. 11, available at www.isee2010.org/paper/23ps0202%23Journey%20to%20world%20top%20emitter%23_Guan,Dabo_.pdf (accessed 11 July 2012).

39 Nina Chestney, "China's CO_2 Emissions Rose 10 percent in 2010: BP Data", *Reuters*, 8 June 2011.

40 Paul Collier, *The Bottom Billion: Why the Poorest Countries are Failing and What Can Be Done About It* (New York: Oxford University Press, 2008), p. 42; "São Tomé and Príncipe: Country Profile", *BBC News*, August 2010.

41 Editorial, "Oil Drilling: ExxonMobil's Big Discovery in Gulf", *Christian Science Monitor*, Associated Press, 8 June 2011, available at www.csmonitor.com/Business/Latest-News-Wires/2011/0608/Oil-drilling-ExxonMobil-s-big-discovery-in-Gulf (accessed 11 July 2012).

42 Michael Watts, "Resource Curse? Governmentality, Oil and Power in the Niger Delta, Nigeria", *Geopolitics*, Vol. 9, No. 1 (2004), p. 54; Thomas L. Friedman, "The First Law of Petropolitics", *Foreign Policy* (Washington), 25 April 2006.

43 Kim Woodard, *International Energy Relations of China* (Stanford, CA: Stanford University Press, 1981).
44 Petropolitics differs from petrocapitalism in that political forces are assessed and hashed out before, during and after petroleum-specific capital investments take place, and that resource-poor neighbouring countries of oil-rich landlocked countries can also play pivotal roles in defining the dynamics of regional political discourses.
45 S. Dalby, "Imperialism, Domination, Culture: The Continued Relevance of Critical Geopolitics", *Geopolitics*, Vol. 13, No. 3 (2008), p. 430; Bill McKibben, *Deep Economy: Economics as if the World Mattered* (Oxford: Oneworld, 2007).
46 José Goldemberg, "Ethanol for a Sustainable Energy Future", *Science*, Vol. 315, No. 5813 (2007), pp. 808–10.
47 Chen Aizhu and Judy Hua, "China June Oil Demand Wanes, at 20-month Low", *Reuters (Beijing)*, 13 July 2012.
48 Amy Myers Jaffe and Steven W. Lewis, "Beijing's Oil Diplomacy", *Survival*, Vol. 44, No. 1 (Spring 2002), pp. 115–34.
49 Information reported by the UN Security Council: "China and Russia Veto US/UK-backed Security Council Draft Resolution on Myanmar", *UN News Centre*, 12 January 2007, www.un.org/news/.
50 In relation to these issues, economic geographers advocate for a redirected focus away from concerns relevant only to the West and the discarding of analytical frameworks that bracket China off from Western experiences. See: Michael Webber, "Re-emerging China and Consequences for Economic Geography", *Eurasian Geography and Economics*, Vol. 51, No. 5 (2010), pp. 583–99.
51 International Energy Agency (IEA), *IEA Response System for Oil Supply Emergencies*, p. 1, 2010, available at www.iea.org/textbase/nppdf/free/response_system.pdf (accessed 11 July 2012).
52 Andrew Batson, "China Seeks More Involvement and More Clout", *The Wall Street Journal*, 31 March 2009.
53 David Barboza, "In Chinese Factories, Lost Fingers and Low Pay", *The New York Times*, 5 January 2008; "Is China's Cheap-labor Party Over?" *The Week*, 18 June 2010, p. 40.
54 Lim Tai Wei, *Oil and Gas in China: The New Energy Superpower's Relations with its Region* (Singapore: World Scientific Publishing, 2010), p. viii; Terry Macalister, "China Puts £60bn into Energy", *The Guardian*, 5 February 2009.
55 Bo Kong, *China's International Petroleum Policy* (Santa Barbara, CA: Praeger & Co., 2010), p. 116.
56 Richard A. Muller, *Physics for Future Presidents* (New York and London: W. W. Norton, 2008), pp. 88–89.
57 This results from dividing the cost of gasoline by the cost of coal, that is, 11 cents by 0.4 and 0.8 cents.
58 These figures do not fully account for the extraction, production, distribution and infrastructure costs.
59 Bankers following China's economy note that "the demand for new vehicles and goods will support a 40 percent increase in diesel consumption to 4.8 million barrels a day by 2015 from 2011". See: Chua Baizhen, "China's Oil Demand to Surpass IEA Forecasts, Barclays Says", *Bloomberg News*, 23 November 2011.
60 Robert Slater, *Seizing Power: The Grab for Global Oil Wealth* (Hoboken, NJ: John Wiley & Sons, 2010), p. 106.
61 US Energy Information Administration, "Saudi Arabia: Oil Exports and Shipping", Country Analysis Briefs, available at www.eia.doe.gov/cabs/Saudi_Arabia/OilExports.html (accessed 27 August 2010). Also, to gain a better understanding of the dry numbers behind preferred energy sources, see: Muller, *Physics for Future Presidents*.

62 Friedman, *Hot, Flat and Crowded*, p. 350; "China's Electricity Consumption up 6%", *The China Post* (Beijing), 7 January 2010; "China's Electric Consumption up in July", *China Daily* (Beijing), 14 August 2010.
63 US Energy Information Administration, "China: Coal", Country Analysis Brief, available at www.eia.doe.gov/cabs/China/Coal.html (accessed 28 August 2010); Also, see: Muller, *Physics for Future Presidents*, p. 300.
64 Elizabeth Economy, *The River Runs Black: The Environmental Challenges to China's Future* (Ithaca, NY: Cornell University Press, 2004), p. 208; Mukand S. Babel and Shahriar M. Wahid, *Freshwater Under Threat – Southeast Asia*, UNEP-AIT Report (Nairobi: United Nations Environment Programme, 2009), pp. 1–44.
65 Elizabeth C. Economy and Kenneth Lieberthal, "Scorched Earth: Will Environmental Risks in China Overwhelm its Opportunities?" *Harvard Business Review* (June 2007), p. 12.
66 For more information, see: United Nations Framework Convention on Climate Change at available at www.unfccc.int.
67 Ernst Hans Gombrich, *A Little History of the World* (New Haven, CT and London: Yale University Press, 2008).
68 For an excellent review on the subject, see: Philippe Le Billon, "The Political Ecology of War: Natural Resources and Armed Conflicts", *Political Geography*, Vol. 20 (2001), pp. 561–84; W. Lei and S. Qinyu, "Will China Go to War Over Oil?" *Far Eastern Economic Review*, Vol. 169, No. 3 (2006); Also, Muller, *Physics for Future Presidents*, p. 63.
69 Ian Taylor, "China's Oil Diplomacy in Africa", *International Affairs*, Vol. 82, No. 5 (2006), pp. 937–59.
70 Pablo Bustelo, *China and the Geopolitics of Oil in the Asian Pacific Region* (Madrid: Real Instituto Elcano de Estudios Internacionales y Estrategicos, Nov. 2005), pp. 2–3.
71 Janet Xuanli Liao, "A Silk Road for Oil: Sino-Kazakh Energy Diplomacy", *The Brown Journal of World Affairs*, Vol. XII, No. 2 (2006); "China, Thirsty for Oil, Looks to Central Asian Neighbours", *People's Daily Online*, 20 April 2010.
72 Juan Forero, "China's Oil Diplomacy in Latin America", *The New York Times*, 1 March 2005.
73 Sigfrido Burgos Cáceres and Sophal Ear, "China's Strategic Interests in Cambodia: Influence and Resources", *Asian Survey*, Vol. 50. No. 3 (2010), pp. 615–39; Tom Bradley, "China's Cambodian Hegemony", *The Diplomat*, 7 May 2009.
74 Fiona Venn, *Oil Diplomacy in the Twentieth Century* (New York and London: Palgrave Macmillan, 1986).
75 Bobo Lo, *Axis of Convenience: Moscow, Beijing, and the New Geopolitics* (Washington, DC: Brookings Institution Press, 2008); "China, Russia Strike $25 billion Oil Pact", *The Wall Street Journal*, 18 February 2009.
76 Mamdouh G. Salameh, "China's Global Oil Diplomacy: Benign or Hostile?" (Cleveland, OH: International Association for Energy Economics, first quarter 2010), pp. 21–25, www.iaee.org.
77 Clifford Krauss and Elisabeth Rosenthal, "Reliance on Oil Sands Grows Despite Environmental Risks", *The New York Times*, 18 May 2010; Jeffrey Jones, "Sinopec to Pay $4.65 billion in Oil Sand Deal", *Reuters*, 12 April 2010; S. Goldenberg, "Canada Looks to China to Exploit Oil Sands Rejected by US", *The Guardian*, 14 February 2010.
78 Thomas Lum, *China's Assistance and Government Sponsored Investments Activities in Africa, Latin America, and Southeast Asia* (Washington, DC: Congressional Research Service, Nov. 2009), pp. 12–14.
79 Simon Romero, "Chavez Says China to Lend Venezuela $20 billion", *The New York Times*, 18 April 2010.

80 Lum, *China's Assistance and Government Sponsored Investments Activities*, Table 8, p. 14.
81 For figures from main crude oil exporters, see: Erica S. Downs, "China", Energy Security Series at Brookings Foreign Policy Studies (Washington, DC: Brookings Institution, Dec. 2006).
82 Peter S. Goodman, "China Invests Heavily in Sudan's Oil Industry", *The Washington Post*, 23 December 2004.
83 Sigfrido Burgos Cáceres and Sophal Ear, "China's Oil Hunger in Angola: History and Perspective", *Journal of Contemporary China*; "Rising Angola: Oil, Glorious Oil", *The Economist*, 28 January 2010.
84 Tom Burgis, "China in $23bn Nigerian Oil Deal", *Financial Times*, 15 May 2010.
85 Marcus Power and Giles Mohan, "Towards a Critical Geopolitics of China's Engagement with African Development", *Geopolitics*, Vol. 15, No. 3 (2010), pp. 462–95, p. 462.
86 Lucy Ash, "China in Africa: Developing Ties", *BBC News*, 4 December 2007; James Traub, "China's African Adventure", *The New York Times*, 19 November 2006; Zhiqun Zhu, "Petroleum and Power: China, the Middle East, and the United States", *Yale Journal of International Affairs* (Spring–Summer 2007), p. 28.
87 Padraig Carmodya and Ian Taylor, "Flexigemony and Force in China's Resource Diplomacy in Africa: Sudan and Zambia Compared", *Geopolitics*, Vol. 15, No. 3 (2010), pp. 496–515, p. 496.
88 Deborah Brautigam, *The Dragon's Gift: The Real Story of China in Africa* (New York: Oxford University Press, 2009), p. 311; Jared Diamond, *Collapse: How Societies Choose to Fail and Survive* (London: Penguin, 2006), p. 9.
89 Marcus Power, "Geopolitics and Development: An Introduction", *Geopolitics*, Vol. 15, No. 3 (2010), pp. 433–40, p. 439.
90 For an excellent review of China's position in central Asian geopolitics, see: J. Richard Walsh, "China and the New Geopolitics of Central Asia", *Asian Survey*, Vol. 33, No. 3 (1993), pp. 272–84.
91 Michael Richardson, "Another 'Pearl' in Beijing's String of Ports", *The Japan Times*, 19 August 2010; Hongyi Lai, "China Builds Bridges to Fuel its Engine Room", *Financial Times*, 4 July 2010, available at www.ft.com/intl/cms/s/0/6e350266-879a-11df-9f37-00144feabdc0.html#axzz20F1K01pP (accessed 11 July 2012).
92 Hongyi Lai, "China Builds Bridges to Fuel its Engine Room", *Financial Times*, 4 July 2010.
93 Hongyi Harry Lai, "China Builds Bridges to Fuel its Engine Room", *Financial Times*, 4 July 2010.
94 Mojgan Djamarani, "Scramble for Africa", *Petroleum Review* (2008), p. 30. Also, more detailed information about this topic can be found in Suisheng Zhao, "China's Global Search for Energy Security: Cooperation and Competition in Asia-Pacific", *Journal of Contemporary China*, Vol.17, N.55 (May 2008), p. 208; Heinrich Kreft, "China's Quest for Energy", *Hoover Policy Review*, 139 (Oct.–Nov. 2006), www.hoover.org/publications/policy-review.
95 Erica Strecker Downs, *China's Quest for Energy Security* (Santa Monica: RAND Corporation, 2006), p. 54.
96 Bo Kong, *China's International Petroleum Policy* (Santa Barbara, CA: Praeger & Co., 2010), p. 116; Michal Meidan, "Shaping China's Energy Policy", *Journal of Contemporary China*, Vol.18, N.61 (2009), pp. 603–4.
97 China's growing presence and stronger economic weight in Asia and the Pacific provide fertile grounds for Beijing to propose accretive bilateral accords and to petition participative membership into regional multilateral organizations. Given that Asian countries now have an option outside of allegiance to the USA and the EU, Western policymakers need to be more creative (and balanced) in their incentive packages should they wish to compete.

98 CBO Financial, *CBO Proves That—Contrary to Popular Belief—it is Easy Being Green* (USA: Algae Technology, CBO), www.cbofinancial.com/downloads/CBO_Algae_Overview.pdf (accessed July 2012).

99 For many examples, see: Thomas Schueneman, "Memo Reveals China Fears Developed Nation 'Conspiracy' at COP15 Climate Conference", available at www.globalwarmingisreal.com/blog/2010/02/22/memo-reveals-china-fears-developed-nation-conspiracy-at-cop15/ (accessed 20 September 2010); Liberal Conspiracy, "Does China Threatens the Environment?" available at liberalconspiracy.org/2010/08/04/does-china-threaten-the-environment/ (accessed 20 September 2010); J. Watts, D. Carrington and S. Goldenberg, "Climate Conspiracy", *Guardian*, 11 February 2010.

100 J. Mitchell, *The New Oil Economy: Impacts on Business, Geopolitics and Society*, (London: Earthscan, 2001).

101 Philip Andrews-Speed, *Energy Policy and Regulation in the People's Republic of China* (The Hague: Kluwer Law International, 2004), p. 339; for a review of multidimensional impacts, see: Mitchell, *The New Oil Economy.*

102 Collier, *The Bottom Billion*, p. 86; Jeffrey Sachs, *The End of Poverty* (New York: Penguin Press, 2005), p. 149.

103 Se Hyun Ahn, "Framing Energy Security Between Russia and South Korea? Progress, Problems and Prospects", *Asian Survey*, Vol. 50, No. 3 (May–June 2010), pp. 591–92.

104 For more on ASEAN, see www.aseansec.org (accessed 20 July 2012).

105 For more on SAARC, see www.saarc-sec.org (accessed 20 July 2012).

106 For more on APEC, see www.apec.org (accessed 20 July 2012).

107 Paul Collier and David Dollar, *Globalization, Growth and Poverty: Building an Inclusive World Economy* (New York: Oxford University Press, 2002), p. 153; similar narratives can be found in Collier, *The Bottom Billion.*

108 Bernard Hoekman and Susan Prowse, "Economic Policy Responses to Preference Erosion: From Trade as Aid to Aid for Trade", World Bank Policy Research Working Paper No. 3721 (Washington, DC: World Bank, September 2005); T. Lloyd, "Does Aid Create Trade? An Investigation for European Donors and African Recipients", *European Journal of Development Research*, Vol. 12 (2000), pp. 107–23.

109 Adherents to neo-colonialism "contend that multinational corporations continue to exploit the resources of post-colonial states, and that this economic control inherent in neo-colonialism is akin to the classical, European colonialism practiced from the 16th to the 20th centuries. In broader usage, neo-colonialism may simply refer to the involvement of powerful countries in the affairs of less powerful countries; this is especially relevant in Africa and Latin America". Arvind Dhond, "Indian Labour Market and Employment: Changing Scenario in Post-Liberalisation Era", *Journal of Research in Arts & Education*, Vol. 1, No. 1 (2012), p. 41, available at www.abhinavjournal.com/images/Arts_&_Education/Jan12/5-Dhond.pdf.

110 Paul Collier, "Is Aid Oil? An Analysis of Whether Africa Can Absorb More Aid", *World Development*, Vol. 34, No. 9 (2006), pp. 1482–83; Also, Paul Collier and David Dollar, *Globalization, Growth and Poverty: Building an Inclusive World Economy* (New York: Oxford University Press, 2002).

111 P. Zeihan, "China: Crunch Time", *Stratfor Global Intelligence* (Austin), 30 March 2010, www.stratfor.com.

112 US Energy Information Administration, "China: Electricity", Country Analysis Briefs, www.eia.doe.gov/cabs/China/Electricity.htm (accessed 28 August 2010); also, see: Friedman, *Hot, Flat and Crowded*, p. 349. Trucks loaded with coal from Inner Mongolia were stuck bumper to bumper for 60 miles for 10 days in late August and early September 2010 in what appeared to be world's biggest and longest traffic jam.

113 For details on the application of energy consumption models, see: Paul Crompton and Yanrui Wu, "Energy Consumption in China: Past Trends and Future Directions", *Energy Economics*, Vol. 27 (2005), pp. 195–208.

114 China energy consumption may have risen 6% to 3.45 billion metric tons of coal equivalent by the end of 2011.
115 J. Fallows, "China's Silver Lining", *The Atlantic* (Washington, DC), June 2008, www.theatlantic.com.
116 Anthony Saichs, *Governance and Politics of China*, 2nd edition (New York and London: Palgrave Macmillan, 2004), p. 97; Lucian W. Pye, *The Dynamic of Chinese Politics* (Santa Monica, CA: RAND Corporation, 1981).
117 Liao Ran, a China specialist with Transparency International says that "the cost of corruption amounts to about 10 to 13 percent of China's annual GDP", as quoted by Andrew C. Schneider, "Seven Flaws in China's Growth Model", *The Kiplinger Letter*, 23 August 2010.
118 China's Foreign Ministry officials have asked the USA to stop the domineering behaviour of exploiting human rights to interfere in the internal affairs of other countries. The USA ignores its own severe human rights problems, ardently promoting its so-called human rights diplomacy, treating human rights as a political tool to vilify other countries and to advance its own strategic interests. See: Chris Buckley, "China Tells U.S. to Quit as Human Rights Judge", *Reuters*, 10 April 2011.
119 Kerry Brown, "China: Between Global Responsibilities and Internal Transitions", in Robin Niblett (ed.), *America and a Changed World: A Question of Leadership* (Hoboken, NJ and London: Wiley-Blackwell and Chatham House).
120 Annette Bohr, "Central Asia: Responding to the Multi-vectoring Game", in Robin Niblett (ed.), *America and a Changed World: A Question of Leadership* (Hoboken, NJ and London: Wiley-Blackwell and Chatham House), p. 121.
121 On this topic, Chinese intellectuals draw inspiration from ancient Chinese philosophy, which regards both material capability and morality as necessary conditions for building strong and durable global leadership. For the sake of making itself a rising power that is welcomed by the rest of the world, China should act as a humane authority and take on more international responsibilities to improve its strategic credibility. Yan Xuetong, "How Assertive Should a Great Power Be?" *The New York Times*, 31 March 2011.
122 K. Hwang, *Face and Favour: Chinese Power Game* (Chicago, IL: University of Chicago Press, 1987); Collier, *The Bottom Billion*, p. 146.
123 "CAS is the national academy for the natural sciences of the People's Republic of China. It is an influential institution of the State Council of China. It is headquartered in Beijing, with institutes all over the People's Republic of China", available at en.wikipedia.org/wiki/Chinese_Academy_of_Sciences (accessed 11 July 2012).
124 Both quotations from Guangding Liu, Changchun Yang, Tianyao Hao and Xiaorong Luo, *Oil and Gas Resources in China: A Roadmap to 2050* (New York: Springer Publishing Company, 2010), p. xvi.
125 Although not specifically on geopolitics, this book presents a wealth of data on resources usage. See: Tristram Stuart, *Waste: Uncovering the Global Food Scandal* (New York and London: W. W. Norton, 2009), pp. 182–83.
126 Recently, China's Commerce Ministry assured the international community that Beijing "will steadily push its long time policy of "go abroad" despite risks caused by the financial crisis". Senior officials claim that "opportunities to invest in high-quality enterprises and assets have increased, investment costs have declined, and transaction conditions have improved". Michael Wei and Jacqueline Wong, "China to Push Ahead With 'Go Abroad' Policy", *Reuters*, November 2010.
127 UNIDO, "Public Goods for Economic Development", UN Report (Vienna, Austria: United Nations Industrial Development Organization, 2008), p. 10.
128 Renyi Zhang, Guohui Li, Jiwen Fan, Dong L. Wu and Mario J. Molina, "Intensification of Pacific Storm Track Linked to Asian Pollution", *Proceedings of the National Academy of Sciences of the USA*, Vol. 104, No. 13 (2007), p. 5298.

129 John Swenson-Wright, "East Asia: Searching for Consistency", in Robin Niblett (ed.), *America and a Changed World: A Question of Leadership* (Hoboken, NJ and London: Wiley-Blackwell and Chatham House, 2010), p. 71.

130 According to a report, the halting of the PLA's naval expansion is exaggerated. Any lull in development is likely to prove temporary. Once the PLA Navy finishes shedding old assets, the submarine fleet will resume its upward trajectory. Estimates indicate that the navy will add 10 modern *Song*- and *Yuan*-class diesel subs by 2015 and an additional 10 by 2020. If such projections are accurate, the fleet will be 78 boats strong. See: James Holmes and Toshi Yoshihara, "China's Naval Build-Up Not Over", *The Diplomat*, 29 September 2010.

131 Viola Gienger and Tony Capaccio, "China Seeks Military Power in Indian Ocean, Western Pacific, Pentagon Says", *Bloomberg News* (New York), 17 August 2010; Zbigniew Brzezinski, "From Hope to Audacity: Appraising Obama's Foreign Policy", *Foreign Affairs*, Vol. 88, No. 1 (Jan.–Feb. 2010), www.foreingaffairs.com; for an elegant discussion on this topic, also see: John Swenson-Wright, "East Asia: Searching for Consistency", in Robin Niblett (ed.), *America and a Changed World: A Question of Leadership* (Hoboken, NJ and London: Wiley-Blackwell and Chatham House, 2010), pp. 71–88.

132 "By the end of 2009, China had provided 256.29 billion yuan (or US$39.27 billion) in aid to foreign countries, including 106.2 billion yuan in grants, 76.54 billion yuan in interest-free loans, and 73.55 billion yuan in concession loans. The Chinese aid went to 161 countries and more than 30 international and regional organizations". Chinese Government's Official Web Portal, "Full Text: China's Foreign Aid", available at english.gov.cn/official/2011-04/21/content_1849913_4.htm (accessed 12 September 2012). Since 2004, China's budgeted foreign aid has increased at an annual rate of 29.4%. Beijing is making headways in foreign aid.

133 According to a White Paper on China's Foreign Aid released by the State Council Information Office, the Vice-Minister of Commerce Fu Ziying noted that "China does not attach any political strings to its aid. Our foreign aid programs are based on the principles of equality, mutual benefit and mutual development. Many developing countries lack hospitals and roads. Our aid is concentrated on sectors where they need it most". See: "China's Foreign Aid Comes with No Strings Attached", *People's Daily Online* (English), 27 April 2011.

134 Chinese Premier Zhou Enlai "proposed the Five Principles of Peaceful Coexistence in 1953 during negotiations with India over Tibet. One major consequence of this policy was that Beijing would not support Communist insurgencies in Southeast Asia, particularly in Indonesia and Malaysia, and would distance itself from overseas Chinese in those nations". See: www.nationmaster.com (accessed July 2012).

II

Country case studies

3 Angola

The continued growth of American and European energy demands coupled with the remarkable rise of Brazil, Russia, China, India and other emerging market economies as major economic players are rapidly moulding international energy (gas and oil) markets. Increasing instability and geopolitical threats in key oil-producing and oil-exporting regions are also contributing to these ongoing market changes.[1]

Evidently, nation-states cannot only be concerned with energy markets changing at a rapid pace, but also with diverse and newer threats that need to be carefully considered alongside the rapid evolution of other components of the global architecture. Some of these components are climate change, public health, education, food security, terrorism, poverty and aid, as well as emergency responses to unexpected natural disasters (i.e. earthquakes, tsunami, volcanic eruptions, etc.) that compete for limited economic, human and physical resources.[2]

For its part, China seems to be concerned with very many events at once. Most, if not all, of these events carry important, long-term strategic implications for the country. In the past decade, Beijing has juggled with multiple contemporary national challenges. These challenges include, but are not limited to, rapid economic growth interrupted by a global recession, rising domestic purchasing capacity, swelling foreign demand for top-quality goods, "loss of energy self-sufficiency, substantial increases in the volume and cost of oil imports", heightened local, regional and international scrutiny of its actions in foreign affairs, social unrest in Tibet, global notoriety owing to both its achievements and setbacks, looming environmental degradation and the associated evolution of policy frameworks that come with responsible statesmanship, among many others.[3]

In relation to dynamic swings in global energy sectors, such as demand-side crude oil price gyrations, China has had its fair share of influence. This can be partly explained by Beijing's purposeful intent on securing, across the globe, the natural resources and raw materials needed to sustain fast-paced economic growth in its manufacturing and industrial sectors. Among the many critical resources needed to fulfil its energy needs, oil is by far its most important, given that it fuels its economic engine. Just as the West looked abroad during its industrial and manufacturing expansion to satisfy its sprouting energy needs, so is China rightfully determined in pursuing diverse energy sources through multiple market-oriented strategies.

The extent of China's ascending oil needs has compelled the country to seek and secure new oil supplies. As part of these seek-and-find efforts, China has turned to the African continent, a vast area of land hosting large oil deposits, a place whose burdens, risks and challenges have often caused it to be shunned and overlooked by the rest of the world. Despite these obstacles, Africa has emerged as the centrepiece of China's overseas energy strategy.[4]

The magnitude of "China's trade with Africa has increased" dramatically in the last 12 years. An idea of the magnitude of annual trade growth follows: $2 billion in 1999; $4.7 in 2000; $9 billion in 2002; $18.5 billion in 2003; $29.6 billion in 2004; $39.7 billion in 2005; $73 billion in 2007; and $106.8 billion in 2008, with a large percentage of "Africa's exports to China arising from oil-rich countries".[5] Using the latest trade value as reference, Sino–African trade has multiplied 53.4 times since 1999, and 3.6 times since 2004.

In practice, Chinese foreign policy towards Africa is driven by the need to secure natural resources and raw materials, a natural result of commercial interests.[6] Throughout the continent, Beijing has conveniently developed tight links and shared interests with a number of dictatorships and repressive regimes. Numerous troubling examples throughout Africa demonstrate that China's rising demand for resources trumps any concerns regarding bad governance, human rights violations and ecological disruptions.[7]

Under the conspicuous notions of sovereignty protection and non-interference that have so often filled Chinese communiqués and public discourses, Beijing and selected Chinese energy companies active in Africa have indirectly legitimized rule-of-law dismissals, undemocratic practices and many civil rights abuses. Although these actions are indeed discouraged and condemned in international arenas, China is not particularly unique in setting up deals with wretched, bellicose governments. The USA and Europe, for instance, are quite prone to negotiating generous deals with recurrent human rights offenders and environmental health violators such as Afghanistan, Cambodia, Egypt, Pakistan, Saudi Arabia and Vietnam.

Irrespective of details, it is clear that China's position of openly acknowledging, without recrimination, all of the violations of standard democratic norms by some African leaders goes against international consensus favouring of fair and responsible statesmanship. In general, China does not seem to regard the political instability, ethnic schisms, or social discords in some African countries as an obstacle to its far-reaching ambitions. Far from it, China and Chinese companies are rapidly seizing burgeoning opportunities and potential deals as they arise, especially as countries start to open their economies to foreign investments and privatization of public services and industries.[8]

This selective opportunism has been highlighted by staffers at the Council on Foreign Relations (CFR) in New York, who ascertain that China is following a pre-established path traditionally followed by the USA, Europe, Australia and Japan of offering developing countries comprehensive technical assistance and aid agreements combined with exploitative trade deals.[9] This comes as no surprise since China, as well as other nations, acts diligently to uphold best the interests of its constituencies.

To put in perspective China's resource hunger around the globe in the context of regional energy dynamics, we must first talk about the progression of demand and supply over time. For instance, Asia's growth in general—and China specifically—is visibly reflected in rising energy and oil demand, whereas gas and oil production in the Asian region has not kept up, supplying less than one-third of Asian regional consumption by 2008. It is estimated that within two decades (i.e. by 2030), the Asia-Pacific oil deficit will be around 70% of consumption, compared to 10% in the Atlantic regions. At that future time, too, it is believed that West Africa could feasibly cover 25% of Asia's entire regional oil needs.

In view of growing demand surpluses and supply deficits in the Asian region, high-level policymakers in Beijing are carefully weighing their options in rapidly changing contexts regarding energy availability in order to uphold national energy security in the medium and long terms. Clearly, Chinese sensitivities to and situational awareness of facts and rumours in relation to regional and global energy determinants are not entirely synchronized with on-the-ground realities. It is for this reason that informed decision making is performed carefully under a case-by-case basis, with implementation of energy and foreign policies proceeding under a revisionist stance that reassesses information as it is gathered, crosschecked and filtered by embassies, representatives, government officials and state-controlled companies in the locations in question.

In practice, and in consideration of the rapidly rising needs aforementioned, oil imports to China, Indonesia and India are mainly conducted by state-owned or state-controlled companies, which actually adjust to the needs of the state much more so than to the whims of corporate executives and foreign shareholders, seeking to maximize quarterly profits.

Big oil suppliers negotiate strategically with big oil importers: some countries peg commodity exports to other reciprocal benefits such as low-interest loans, migrant hosting, aid assistance and commercial leniency. Middle Eastern exporters, for instance, go to great lengths to ensure that contract resell prohibition clauses are strictly enforced so that no unnecessary friction arises. These clauses are necessary to avoid having rogue officials in top government positions conducting "buy low, sell high" transactions with other countries that do not enjoy "special" treatments, as China, India and others do. Furthermore, this hawkish oversight by Middle Eastern oil companies results from the sheer size of their export—that is, the supply of 70% of their oil outputs to the Asia-Pacific region in 2008.

The classical balance of power in oil rested largely on Middle Eastern countries, but recently a new and evolving oil axis has emerged. The term "oil axis" has been used to describe a number of oil-dependent relationships, strategic partnerships or possible coordination between countries in their quest to secure external energy resources or to contest hegemonic positions. The new oil axis we refer to includes dominant players such as China, Russia, Iran, India and Venezuela. For example, the most strategically significant manifestation of this oil axis is Russia's willingness to use its leveraged resource position to counteract what Moscow considers an unacceptable level of American infringement on its territorial interests and spheres of dominance. Venezuela, too, tries to undermine US influence in Latin America.[10]

In view of the aforementioned evolving oil axis, selective investments in oil-containing zones have a commercial strategic value and national security interests. Here, dominant countries and giant oil companies compete ferociously for the most valuable and reliable options, while seeking continued support from local governments interested in securing oil export revenues to finance military spending, domestic growth, or development.[11] In Africa, China chose resource-rich Angola.

Brief outline of this chapter

This chapter makes an attempt to lay down the most relevant and critical pieces of information that serve better to demonstrate and explain China's oil hunger in Angola. First, it presents Angola's position in terms of its economy, politics and resource wealth. It explains why Angola has positioned itself as a pre-eminent crude oil provider to China. Second, it presents China's growing resource needs, especially that of oil and natural gas, as well as extending its analysis on the breadth and depth of strategic approaches used by Beijing to accomplish its objectives. A conclusion offers summary findings and guidance on the way forward for both countries.

Angola's oil richness

An idea of the magnitude of Angola's oil richness and its resource dependency can be obtained by looking at these key figures: crude oil accounts for 90%–95% of export earnings, 80% of government revenue, and more than 50% of its gross domestic product (GDP). This oil-fuelled windfall has been growing steadily over the past three decades as a result of higher crude oil outputs coupled with rising oil prices.[12]

Natural resource explorations in the 1950s led to the discovery of vast oil deposits throughout Angola's territory. As of early 2010, Angola's oil reserves were estimated to be between 12 billion and 14 billion barrels, with an approximate daily output of 2 million barrels. If oil reserve estimations are accurate and by keeping current daily output rates fixed into the future, its deposits are projected to last anywhere from 16.4 to 19.2 years.

Since 1968, once early production from shallow waters started off the coast of the northern province of Cabinda, Angolan authorities intelligently pursued a policy of strategic diversification in its energy partnerships; however, not very many partnerships followed given that Angolan development was interrupted by almost half a century of strife, political contention and popular unrest.

From 1970 until 1975, Angola waged a war of independence against its former colonial masters, the Portuguese. This period was "followed by nearly three decades of violent fighting between the belligerently communist Popular Movement for the Liberation of Angola (MPLA) and the pro-Western National Union for the Total Independence of Angola (UNITA) that ended in 2002. Before the 1970s it was virtually impossible to launch expansive and uplifting entrepreneurial initiatives like the one currently witnessed given that Angola has had a long history of armed conflict".[13]

Angola's struggle was so distorting, convoluted and longwinded that by the end of it, primary and secondary infrastructure along with political groups, judicial frameworks, governance, "institutions and social services had to be revamped or rebuilt from scratch".[14]

Throughout the civil war and beyond, oil was the main driving force attracting most investments into the country. By 11 November 1975, Angola's independence was attained, and one year later, the National State Oil Company of Angola (Sonangol) was established under Decree 52/76 as a state-owned entity with a mission to manage hydrocarbon resource explorations in Angola. By 1991, Sonangol was attributed the first deepwater oil concession.

Once political bickering was put to rest, sustained peace and bankable democracy was finally achieved in 2003, which brought the political stability necessary to draw in investors and investment confidence, and attracted the required financing capacities to develop their nascent oil industry. A few months later, by early 2004, Sonangol began full operations as a formal oil company in several offshore locations, and "opened its SonAsia office in Singapore, aimed at promoting the trade of Angolan crude oil to Asia".[15]

In subsequent years, following the examples of Nigeria and Côte d'Ivoire, the Angolan oil ministry, with inputs from power elites, divided its continental shelf into 35 separate blocks—commonly known in the industry as deepwater oil fields—which were auctioned to gain the "rights to develop extraction activities in partnership with Sonangol. Most of these have been offered to international oil companies", such as ExxonMobil, which is currently the largest American operator in Angola's oil industry. Additionally, four other Western oil companies are building up substantial investments in Angolan oil blocks, with little competition from Asian firms because of the complexities of managing deepwater oil fields.[16]

Aided by a much less volatile security environment, and with government support behind a state-owned entity, the extent of domestic and foreign oil-related investments and energy-securing commitments is further evidenced by the astounding rise of crude oil production. It started rather weakly from roughly 170,000 barrels per day in 1975, more than quadrupling to 800,000 barrels in 2002. The galloping pace of infrastructure progress and up-scaled production capacity in the last seven years has resulted in a dramatic jump in crude oil output from fewer than 1 million barrels in 2003 up to 1.8 million barrels by 2009. Currently, production is set at 2 million barrels per day, but it is expected that production will reach 2.25 million barrels per day by the end of fall 2010. Angolan economists and bankers say that in 2006 and 2007, real GDP growth reached just about 20%, mostly explained by revenues derived from skyrocketing oil exports.[17]

Conjointly, Angola's vast oil reserves and flexible production levels make it a leading producer and exporter of crude oil. It is, therefore, not at all surprising that Angola quickly became a "strategic oil supplier to the world's largest oil consumers".[18] As a result of supply-side market stimulation, between 2004 and 2009 Angola posted record crude oil production levels, ahead of oil outputs reported by Azerbaijan, Brazil, Kazakhstan, Libya and the Russian Federation. By 2007 Angola supplied 29% of US oil imports, and by mid-2008 it managed to "surpass Nigeria as the leading sub-Saharan crude oil producer",[19] and by early 2009, it was the second largest oil exporter to China and the fifth largest to the USA. Such are the country's stockpiles that between 2008 and 2009, Manuel Vicente, president of Sonangol, stated that "oil outputs could be easily managed from 1.92 million to 1.656 million barrels per day, in order to comply with the quotas" established by the 12-country[20] Organization of the Petroleum Exporting Countries (OPEC).[21]

On a separate line of discourse, it is worth noting that from 2003 onwards, the sheer speed of structural and economic development in Angola has been simply remarkable. For instance, in Luanda, the capital, as well as in other urban centres throughout the country, an urbanization boom is taking place as roads, highways, ports, railways, office buildings, hotels, shopping centres, hospitals, universities and apartment blocks rise up out of a green-jungle bush environment. Furthermore, its oil-plenty exuberance also has been used to repay debt in oil deliveries instead of cash. However, political scientists and economists believe these in-kind payments are a result of weak governance and foreign suggestions, thus seriously questioning the viability of this practice if oil prices in global markets should suddenly plunge.[22]

Paradoxically, and in line with recent experience, by 2008 oil prices crashed with the global recession,[23] and along followed Angola's oil-dependent economy. It is estimated that in 2009, the Angolan economy grew, at best, by 1.5%. However, with demand for oil slowly rising as reserves dwindle, it is rebounding strongly. Economists in the region believed that "Angola will be among the world's top five performers" in 2010, with growth exceeding 8%; a little behind China, Brazil and Russia, but 3% ahead of the continent's 5% growth rate, according to African Development Bank estimates.[24]

All these ambiguous estimates are at best wishful thinking based on ever-higher demand prospects. Commentators at leading think tanks note that since the start of the global economic recession, oil demand has fallen dramatically at a rate of 0.4 million barrels per day in 2008 and, based on a consensus forecast, by 1.7 million barrels per day in 2009. As a result, global crude production spare capacity swelled nearly 20% by the end of 2009.[25]

Of relevance to Angola's coffers is the fact that oil stocks have risen to unprecedented levels in recent times, thereby creating an over-supplied market. This is bad news for oil-producing countries, since it now seems that the prospects for higher oil prices will depend very much on the speed and strength of global economic recovery. Notwithstanding, there is still hope in the possibility of "global financial funds dumping money into commodities" since government bonds look sadly unattractive given serious budget deficits following expansionary fiscal policies to combat the Great Recession, which, it is hoped, would force oil prices higher.[26]

On another front, it appears that institutional support is building in favour of supporting and maintaining higher oil prices, with the former chairman of OPEC, Ecuador's Germanico Pinto, stating that quota calibrations are conducive to keep prices above $70–$80 per barrel of crude oil,[27] and these price stabilization efforts are further evidenced on volatility charts on a recent oil market report by OPEC.[28]

As noted above, a downward sliding oil price "from an average of nearly $100 a barrel in 2008" to slightly more than $70 by mid-July 2010 has dented Angola's fortunes. Despite this income retreat, it has continually relied on lump-sum loans and "international lines of credit for infrastructure projects".[29] China has taken the lead in providing much of the long-term financing. For instance, since 2002, the Export-Import Bank of China has lent Angola $4.5 billion, and Chinese privately owned oil investment funds have provided another $3 billion, with more disbursements available upon request.

The West, not wanting to lag behind, has also been present in trying to lure Angola into using its international financial institutions. In late 2009 the International Monetary Fund (IMF) provided $1.4 billion as a standby loan to cover multiple expenditures. Brazil, a vibrant emerging economy in its own right, has facilitated a $1.8 billion line of credit, with Portugal adding another $1.4 billion under the same format. Overall, foreign direct investment (FDI) has more than doubled in the last four years. In 2005 FDI reached $6.8 billion, but jumped in 2008 to $15.5 billion as a result of increased demand to develop much of the needed oil and national infrastructure, which has had spill-over effects in downstream oil-associated businesses.

It would be misleading, however, to imply that all these economic blessings have been solely imparted by capital market forces. The truth is that the trend of public-private reconciliation between and within countries has been relentlessly upward, thus aiding the forward-looking process with soft courting of influential players and issue champions. For instance, former Chairman of OPEC José Maria Botelho de Vasconcelos, also oil minister for Angola, noted in a China-sponsored summit that Beijing plays a proactive "role in the economic development of his country", especially in the agricultural, construction and energy sectors,[30] and in early March 2010, João Bernardo,

Angola's ambassador to China, praised the uptick in China-Angola bilateral cooperation.[31]

Members of the opposing political party, UNITA, Angola's intelligentsia, and international observers at diplomatic and United Nations (UN) posts criticize some of Beijing's ways of conducting business in Angola, claiming that in many instances they have been open about their willingness to offer and pay bribes to public officials, and attaching little or no conditions to unilateral loans that undermine local and international efforts to improve public-sector governance and disrupts incentives for much-needed macro-economic reforms by national ministries.[32] The Angolan private sector has also expressed its concerns about Chinese service sector firms not hiring Angolans and underbidding local competing firms.

Despite expressed misgivings and the largely negative portrayal of incoming Chinese investments in Africa, in many quarters their presence is not only welcomed but, nonetheless, cherished.[33] As a goodwill gesture Beijing has unambiguously erased $10 billion in bilateral debts with many African countries, it has also provided academic scholarships for African students to seek higher education in Chinese universities and training centres, and has opened its borders to host hundreds of displaced African workers that have by now joined the ranks of low-cost labour force.[34] In Angola specifically, civil engineers and architects note that buildings and "infrastructures built by Chinese firms are of good quality, low cost, and oftentimes completed in less than half the time such projects would usually take in Angola".[35]

In general, for all the oil Angola has and the development prospect this oil portends, the country is fraught with bureaucracy that impedes progress of almost every economic activity. As with all Africa, Angola is considered to be one of the world's most difficult regions to undertake business. The seaside capital of Luanda, with its port, is an operational nightmare in terms of business logistics. Exports of coffee, cotton, salt, sugar and iron are often delayed. It lacks a skilled workforce that would be required to jump-start the service sector. The port of Luanda, the largest in Angola, is clogged and understaffed, with connecting roads poorly marked, streets to and from nearby cities potholed and jammed with traffic. In addition to this, electricity, required for the creation of a manufacturing base to diversify the economy, is erratic.

Not surprisingly, the effects of "Dutch disease" or the "resource curse" are already apparent, with corruption, industrial stagnation, protectionism and uncompetitive exports leading the way. Despite government budget deficits, taxes remain relatively low, yet obtusely complex to calculate and file, with the highest rate of income tax of 17% and corporate tax of 35%. Weak governance and institutional failures are evidenced by strong patronage networks, manifesting themselves in corruption,[36] mismanagement and nepotism making headlines particularly after oil licence announcements.

In summary, analyses of annual trade statistics suggest that Angola has positioned itself as China's crude oil provider.[37] According to figures recently released by the Chinese Customs Administration, China's year-on-year trade with Angola totalled $2.04 billion in January 2010, or 111% more than in January 2008, with Chinese acquisitions totalling $1.85 billion, which constitutes roughly 91% of total trade and 155% more than in January 2008.[38] Furthermore, "since 2002, Angolan oil exports to China have increased sevenfold, a rate twice that of exports to the United States".[39] Angola's collective production systems are so dynamically vigorous that by January 2010 it became China's top oil supplier, finally overtaking Saudi Arabia from the lead.[40] However, this vigour should be seen in the context that the leading supplier of crude oil to China can change regularly throughout the year, even on a monthly basis. This fierce competition is taking place amid times when Chinese oil companies overseas are vigilantly engaging in geostrategic activism, thereby seeking slowly to consolidate Angolan offshore blocks. In spite of the volatile changes taking place, as of mid-2010, Angola managed to supply 24.84% of China's oil imports.

China's oil hunger: deconstructing consumption from the bottom up

China realized early on that it needed ample access to critical commodities not only to feed an expanding population base, but also to meet the insatiable demands of output and export-oriented industries. Its manufacturing and industrial apparatus is thirsty for natural resources such as forest products (hardwood, softwood and timber), gas, oil, water, rubber, lands, crops, microchips, plastics, diamonds and minerals (aluminium, copper, gold, silver and iron ore).[41] These colossal levels of input requirements are needed to sustain its economic growth prospects.[42] Today, "China relies on coal for most of its energy needs, yet it is the second largest consumer of world oil behind the United States", importing 4.03 million barrels per day (half of which could be supplied by Angolan exports) in January 2010.[43]

A Brookings Institution report notes that by 2005, China's most significant crude oil imports were coming from Saudi Arabia (18%), Angola (14%), Iran (13%), and the Russian Federation (10%), thus reflecting a strategic diversification of supply sources without exclusive or heavy reliance on any particular geographical location.[44] In 2006, the percentages of total imports slightly changed to account for new suppliers, geopolitical interests and rising demands, with oil coming from Saudi Arabia (16.4%), Angola (16.2%), Iran (11.6%) and the Russian Federation (11.0%). It is noteworthy that by late 2006, "China was the second-largest importer of crude oil from Angola after the United States",[45] but in 2007 China overtook the USA as the largest importer of Angolan oil. A year later, it further extended its lead, with Angolan oil accounting for 18.2% of China's total oil imports, more than four percentage points of import volumes from Angola three years before. Ironically, China was once the largest oil exporter in Asia; however, from 1993 onwards it became a net oil importer, and is therefore now dependent on a portfolio of suppliers to satisfy rising needs.[46]

It is important to view the Sino-Angolan connection from the context of Angola's contribution to total Chinese energy draw. Most estimates suggest that one-quarter of China's total energy needs are met by a portfolio of products from imported sources. Of that amount, a little more than half, or about 15%, is met by crude oil. Given that China's total Angolan oil imports currently amount to roughly 25%, then it would suggest that Angola's contribution to China's overall energy draw is 3.75%.

In terms of total volume of crude oil imports and oil consumption, China imported 190 million tonnes in 2007, 198 million tonnes in 2008, and 204 million tonnes in 2009. According to a September 2012 update by the US Energy Information Administration, China consumed an estimated 9.8 million barrels per day of oil in 2011, up 400,000 barrels per day, or over 4% from 9.4 million barrels per day in 2010, with close to 60% used as fuel for air, land and sea transportation; and it is also estimated that by 2020, China may absorb up to 16% of the tradable oil in world oil markets.[47]

"China defines energy security in terms of two issues: price volatility and security of delivery".[48] Moreover, Beijing's energy securement strategy in Africa can be analysed and discussed as a three-step approach. The first step is strong: it aggressively courts and lures the largest oil-rich countries and their governments by offering a sophisticated portfolio of economic diplomacy consisting of aid and development packages, resources-for-infrastructure construction agreements, debt forgiveness, preferential arms sales and non-ammunition trade deals. In Angola, for instance, after a cease-fire was reached between MPLA and UNITA, Beijing flavoured its entry bid by offering a $2 billion package of aid and loans. This was followed with significant resources for Chinese firms to build government office buildings, bridges, hospitals, roads, parks, schools, railroads, information-enabling technologies and fibre-optic telecommunication networks. Foreign journalists and blue-collar Angolans believe that this initial money-laden reconciliation helped China secure a prominent stake in Angola's prospective development future.[49]

The second step is weak: an inconspicuous, silent and almost imperceptible pursuance of consultation, "exploration and oil production agreements in smaller, low-profile countries such as" Côte d'Ivoire, Gabon, the Republic of Congo, Sudan and Equatorial Guinea, with oil deposits "of a size, quantity and quality of little interest to international oil companies"[50] and Western oil giants. This step has proven remarkably successful to Beijing's global energy strategy, since Sudan—for some time one of its key flagship energy investments—managed to export over 60% of its oil outputs to China.

The third and last step is neither strong nor weak, but opportunistically utilitarian: it focuses on oil intelligence—that is, the visioning exercises that systematically scan the African landscapes, hunting for new oil sources and investment opportunities. It makes clever use of Chinese embassies and firms to collect pertinent information about oil deposit discoveries, political turmoil in countries with difficult entry access, problems encountered by oil competitors and pliable collaborators in host governments. Commentators note that Beijing-based think tanks frequently hold stakeholder consultations with senior government and military officials, business leaders and intellectuals to influence policymaking in Beijing, so as to indirectly encourage management groups of "state-controlled oil companies to actively seek out exploration", production and distribution "contracts with countries that produce gas, oil"[51] and non-oil commodities in Africa, but elsewhere too.[52] To Beijing, Angola simply represents a mere tool of material advantage.[53] For more information on actions and strategy, see Chapter 6.

In this context, China's brazen resource hunger has been so widespread that in some instances Chinese companies have come close to taking over entire African economies. For example, Guinea, a West African nation with substantial reserves of bauxite, diamonds, gold, iron and uranium, has seen its fair share of power-grabbing competitions by international companies, jostling control of largely untouched natural resources. Moreover, with the aid of army Captain Moussa Dadis Camarra, former president of Guinea, the Hong Kong-based China International Fund Ltd, along with the Guinean Government formed a Singapore-based firm called the Guinea Development Company (GDC),[54] which managed to seal agreements on a series of deals that would hold the absolute rights to all of the country's gas, oil and minerals except for those already subject to existing contracts. The China International Fund acquiesced to deliver investments in the amount of $7 billion–$9 billion on an economy with a GDP that barely reaches $3 billion.

Chinese interests have been successful in opening further negotiations to buy other privileged Guinean assets, including, as expected, oil exploration rights covering almost all their coastal waters. Similar multi-billion-dollar deals by Chinese enterprises have cropped up elsewhere in Africa, such as in Sudan and Zimbabwe.[55] It is remarkable, none the less, that with significant uncertainties in vulnerable oil markets related to geopolitical disruptions that could derail oil supplies and, therefore, oil prices too, China is moving forward unfettered by vexing world events. Confronted with unknowns, rising incertitude could cause price volatility that could call up decision makers in Beijing to re-evaluate expansionist and interventionist plans.

Unsurprisingly, the same China International Fund Limited (CIFL) that concocted deals in abutting oil-rich countries appeared on the Angolan political and economic scene in 2005 in connection with the design and implementation of infrastructure projects under the auspices of the National Reconstruction Office (Gabinete de Reconstrução Nacional—GRN), which resulted in a "$2.9 billion credit facility to assist Angola with its post-war reconstruction efforts".[56] In the context of corruption allegation and high-level public-private sector reconciliation, it is worth mentioning that the GRN is exclusively overseen by and accountable to the Office of the President of Angola.

CIFL is largely owned by an international development consortium and Sonangol Research and Production Company. It is believed that the fund leverages its governmental partnerships and top-level connections to create a true commercial and industrial behemoth; erstwhile, "key individuals involved with" CIFL may also have "close links to the security apparatus of China".[57]

From the outset, senior Sonangol executives have assiduously focused on diversifying the concerns of business activities to develop international joint ventures and establish related companies that promote both the social development of Angola and the expansion of its corporate mandate. In coupling Sonangol's diversification plans and Beijing's energy securement strategy in Africa, the natural outcome has been a blurring of the lines between public- and private-sector interests. Under this scenario, Chinese executives have been clever about forming different legal entities as limited liability companies to diversify potential fallout and lawsuit risks, and also for more effective managerial purposes. These companies include, but are not limited to, China Sonangol Asia, China Sonangol Engineering and Construction, China Sonangol Exploration and Production, China Sonangol Natural Resources, China Sonangol Finance, China Sonangol Gas, China Sonangol International Investment, most of which are registered in Beijing, Hong Kong and Macau.[58]

On the continent, for example, China Sonangol International Investment, the financing consortium with Africa-wide tentacles, is a joint company that is 70% controlled by Angola's Sonangol and 30% owned by Dayan International Development, a firm that upholds the interests of the China International Fund. These two corporations work in lockstep through African regions by searching, pursuing, negotiating and securing investment opportunities, oftentimes with an upfront no-questions-asked policy.[59]

The extent of involvement of Chinese interests in African territorial and developmental matters is not limited to the western part of the continent or low-income countries. For instance, in 2007 the Tanzania Petroleum Development Corporation reportedly granted three oil licences to China Sonangol International in Lake Rukwa, within the premises of the Uwanda Game Reserve, in the west of the country. Furthermore, in 2008 Tanzanian authorities announced another agreement with China Sonangol International to turn over a minority stake on Air Tanzania, in addition to delivery promises of nine new aircrafts and an extensive renovation project of Dar-es-Salaam airport near the capital city.

That same year, China Sonangol International Investment also managed successfully to acquire deepwater oil mining rights in Nigeria, while, in the Côte d'Ivoire, it reached a financing accord for joint oil exploitation and construction of a refinery with national counterparts.[60] Thus far, the China Petroleum and Chemical Corporation (Sinopec)[61] currently controls more than 5 million tonnes of oil in Western Africa, and since 2004, Beijing has obtained equity partnerships in Angolan deepwater oil fields through Sinopec's majority in Sonangol Sinopec International Ltd—a joint venture between Sinopec and Sonangol—and in the easier access "shallow-water blocks through the China Sonangol International Holding Ltd".,[62] a joint venture between Sonangol and Hong Kong-based private business interests. Conveniently, too, to avoid Nigeria's woes in terms of project-specific financing issues, all Angolan deepwater oil fields are governed by production-sharing contracts. This modality has been increasingly attractive to foreign oil companies as it makes exploration less uncertain and production more reliable.[63]

More recently, Sinopec and China National Offshore Oil Corporation (CNOOC) were in preparations to acquire the 20% stake held by the US-based Marathon Oil Co., but Sonangol opportunistically levied its right of preference in the deal, forcing it to pay over $1 billion to increase its ownership stake to 40%. In the increasingly competitive landscape of world oil markets, Chinese oil firms not only compete with each other for the most profitable and viable projects in Africa,[64] but also with the more politically and technologically shrewd international oil firms in Europe and the USA. Yet, for all the soft power Beijing exerts and its strategic economic diplomacy, it is clear that in some selected cases, China's national oil companies are far more powerful and politically stronger than the specialized government oversight agencies charged with regulating them, domestically and internationally.

This phenomenon applies also to private-sector companies abroad where the regulatory units of the Chinese Ministry of Commerce and the Ministry of Foreign Affairs have little sway and no overarching authority over Chinese private corporations overseas.[65] For all intents and purposes, private- and public-sector energy companies follow the guidelines and prescriptions of the Chinese Ministry of Land and Resources and the recently formed National Energy Commission (NEC), which is "headed by Premier Wen Jiabao, with the Vice-Premier Li Keqiang acting as deputy head".[66] Both entities have authority over state-owned companies when it comes to take charge of China's energy and natural resource policies for better coordination in formulating strategy and planning development.

Hawkish observers in Brussels, London and Washington seem concerned by the growing influence China has on African countries, where Beijing has opportunistically displaced many Western nations with the sheer weight of its crosscutting investments. In response, US Secretary of State Hillary Clinton sought during a visit in mid-2009 to strengthen ties with Angola, trying to improve a relationship that ebbed during the dismissive eight years of the George W. Bush Administration.[67] This renewed reconciliation is not only important, owing to the fact that the USA imports 7%–10% of its oil from Angola, which rivals Nigeria as the continent's biggest oil producer, but also because it seeks greater market shares and investment opportunities for US companies, where it is losing ground.[68]

Over the short term, the Chinese crude oil appetite seems to have been temporarily diminished by the global economic slowdown, in general, and decreased US import volumes of Chinese goods coupled with a domestic retail product sales slump, in particular. Throughout Chinese territories, automobile sales have been cut by one-fifth, and transportation fleets downsized to accommodate depressed demand of manufacturing and industrial products. Politically, Beijing faces its biggest challenge in years as the prospect of economic recovery is extended forward and unemployment rises among its young working class. In response, it is already taking drastic steps to try to limit unemployment by kick-starting domestic infrastructure development projects such as oil and gas pipelines linking Myanmar (Burma) to its southern province of Yunnan, and other alternative energy sources initiatives.[69]

In the end, regardless of geographic focus, China's resource hunger is justified by its right to develop as a nation. During the Industrial Revolution, from the 18th to the 19th centuries, the West embarked on profound agriculture, transport and manufacturing changes that could only have taken place with satiation of rising energy needs to move ahead. In particular, the USA, the UK and other European nations sought a variety of energy sources at home and abroad to fuel their industrial growth. It is to this precedent that China looks, in order to justify its current actions. Although a collective responsibility of prosperity has permeated much of the multilateral discourse fuelled by Western liberalism, the urgency of climatic concerns has turned economic development prospects against emerging economies.

In recent summits, the tone has been apprehensive owing to ideological discord, mainly coming from emerging international players such as Brazil, Russia, India and China, alleging that traditional hegemonic powers are using global contemporary challenges like global warming, terrorism and increasingly fragile security architecture to stop their advance. Indeed, many of these hazardous threats affect the global commons; however, for these countries the heightened primacy of economic development outweighs environmental stewardship or religious wars.

Beijing, as well as Tehran and Moscow, do not necessarily want the unipolar world that Washington protects out of self-interest, which, as expected, invites internecine strife among key players, without which most of the solutions needed could not be attained. As the world's population continues to rise and nears 7 billion, so does the demand for food and energy that is to be supplied in spite of resource limitations and within natural boundaries of living systems.[70] In the moral balance for upholding national interest, the East and West are working against a ticking clock on the edge of global deterioration; thus, it is not uncommon to hear compelling calls for international standards of conduct. Beijing has laid out its strategic roadmap under a revisionist framework that matches functionalities with realities, yet it falls short of feeling guilt or remorse for seeking a prosperous path previously taken by others.

Concluding remarks

Without a doubt, China's tactical diplomacy, soft power and growing economy would allow it to become a top international player, whereas Africa is the world's final frontier, a rare place with ample room to dramatically increase crude oil and agricultural production to meet rising demand around the globe.

If the neglect of Africa's development continues, we all suffer. However, this should not be the case as there are abundant opportunities for corporate investments: just as public and private Chinese companies are active in Angolan territories building roads to transport raw materials needed to fuel their industries and factories back home, the same approach can be followed in other sectors, such as agriculture, manufacturing and services. African governments are, however, wary of a new type of disguised colonization, this time at the hands of transnational corporations. These fears arise as a result of the appalling track record of some foreign oil and mining firms which have extracted vast amounts of raw materials and left little benefit behind.[71]

In the context of potential investment opportunities for Chinese and non-Chinese firms, it is, none the less, important to keep in mind that investors (individual and corporate), as well as sovereign funds, pay particular attention to a number of factors including financial system adequacy, political risk, transparency, the legal and regulatory environments, accountability, foreign exchange risk, governance, oversight styles, brewing rebellions, patronage networks, institutional capacity, and the existing infrastructure to support commercial and industrial activities, to name a few.

China's foreign policy in Africa was designed, and continues to evolve, based on the most pressing economic and strategic considerations at hand; therefore, the subsidiary policy instruments and associated actions are at their very outset downright selfish, opportunistic, utilitarian, non-intrusive and, by default, self-serving. Although non-interference has been heralded as Beijing's assistance motto, it has quickly learned that in some special cases this is not the best path to take—now fully aware that entrusting significant business interests to unscrupulous and repressive governments has its damning perils. However, for all the threats and risk it carries, the notion remains part of its official rhetoric.[72] Moreover, because sustaining economic growth and furthering social well-being are critical goals to advance, Chinese leaders cannot afford the irresponsibility of not securing the country's energy supplies at any possible cost, especially since failing to do so could have an incalculable political price.

As Beijing harmonizes its foreign and energy policies, it should consider the rapid changes taking place in global energy markets. First, "big Western oil companies are trying to expand through investments and acquisitions, but the opportunities are becoming scarcer". Second, and to make matters worse, "state-controlled rivals in the Middle East, Russia, and Central Asia jealously guard oil reserves at home". Third, highly competitive state-run oil companies force transnational "oil giants to look farther afield for new oil reserves", but few new significant oil fields have been discovered in recent years. Fourth, international oil companies now spend significant sums of money on first and "second generation biofuels made from corn, sugar cane, algae or waste materials, but these could take years to develop".[73]

As a whole, dwindling reserves, limited opportunities and fierce competition are driving diversification into other energy options. China should not lag behind. Commentators claim that natural gas is by far the safest bet to make. Indeed, natural gas "is increasingly important for power generation and heating, and the global markets are expected to grow by half by 2030. Big oil companies are keen to expand by building gas-liquefaction plants that make gas readily transportable". However, the strongest incentive still is that world "oil prices, unlike gas prices, are supported by the twelve nation-states that make up OPEC".[74]

Although it is anticipated that Beijing will sometimes fall short of meeting Africa's needs, wants, expectations and whims, it is clear that China's vast government machinery is reassessing, reformulating and adapting its policies and frameworks to align itself with the realities on the ground. This process is far easier in Africa than in Latin America and elsewhere, since the African Union (AU) lacks an official, coherent, guiding policy with respect to China, thereby weakening the entire continent's ability to argue, negotiate and concur with intimidating technocrats in Beijing. However, this is not necessarily a desirable outcome for which all African countries fight, certainly not Angola, given that individual countries benefiting from China's largesse do not want the AU or any other oversight body to interfere in their accretive bilateral dealings.[75]

Whether or not the AU develops and takes a formal stance with regards to Beijing's interventionist agenda is at this point a peripheral outcome. The real thrust lies in encouraging African leaders and their governments to enact sound and rigorous reforms on trade tariffs, production infrastructures, investment and food safety regulations, domestic and international market access, transparency and institutional governance. Doing so would help to fully realize the benefits of having numerous Chinese firms helping African entrepreneurs to gain ample and easy exposure to foreign value chains worldwide so that they can increase with confidence the quantity, quality and diversity of their export products. In practice, for all the positive factors engaging China brings, there are opposite and equal negative factors.

A Nordic research team composed of social scientists says that despite significant advances in market and trade liberalization, international investments will become more difficult in years to come owing to the rising number and magnitude of hidden costs associated with fiercer competitions for limited natural resources in the "global anywhere".[76] Other commentators have gone further in examining the assumptions of sustainable growth and global capitalism as a cause and consequence of cyclical economic crises,[77] with full cost pricing considerations pushing consumer prices ever higher.[78]

It is unjust, however, to narrow China's appetite for resources strictly to oil. Over the years, domestic Chinese firms have been active importers of substantial amount of non-oil commodities in African markets such as, for example, aluminium, copper, cotton, iron, nickel, rubber, spices and timber, thereby lifting many Africans out of poverty. Furthermore, China has begun to import sizeable quantities of value-added, quality-controlled African manufactured goods such as household consumer goods, trinkets and processed foods, among other items.

If a long-term, synergistic and mutually beneficial partnership between Angola and China is envisioned, then a prosperous, vibrant and stable Angola is in line with China's best strategic interests. Given that Beijing's commitments, investments and engagements with Angola will not likely abate any time soon, Washington, London and Brussels should consider a coordinated approach with influential donors to advance a progressive neo-liberal agenda of market openness, economic liberalization, religious freedom, political pluralism, enhanced accountability, social justice and transactional transparency that are intimately linked to and incentivized with foreign aid and technical assistance, which, in turn, will limit China's influence in Angola, but also contribute to international development and nation building.

Acknowledgement

This chapter is adapted from a refereed article entitled "China's Oil Hunger in Angola: History and Perspective", by the authors, published in the *Journal of Contemporary China*, Vol. 21, No. 74 (2012), pp. 351–67.

Notes

1 Scott L. Montgomery, *The Powers That Be: Global Energy for the Twenty-first Century and Beyond* (Chicago, IL and London: University of Chicago Press, July 2010), pp. 1–4.
2 Robin Niblett, *America and a Changed World: A Question of Leadership* (Hoboken, NJ: Wiley-Blackwell, April 2010); and James L. Smith, "World Oil: Market or Mayhem?" *Journal of Economic Perspectives*, Vol. 23, No. 3 (2009), pp. 145–64.
3 Yong Li and Andreas Oberheitmann, "Challenges of Rapid Economic Growth in China: Reconciling Sustainable Energy Use, Environmental Stewardship, and Social Development", *Energy Policy*, Vol. 37, No. 4 (2009), pp. 1412–22, p. 1419; Joanna Lewis, "Climate Change and Security: Examining China's Challenge in a Warming World", *International Affairs*, Vol. 85, No. 6 (2009), pp. 1207–08; "The Outlook for China's Economy", *Forbes* (New York), 8 April 2009.
4 Ian Taylor, "China's Oil Diplomacy in Africa", *International Affairs*, Vol. 82, No. 5 (2006), pp. 937–59.
5 Christopher Alessi and Stephanie Hanson, "Expanding China-Africa Oil Ties", backgrounder (New York and Washington, DC: Council on Foreign Relations, February 2012), available at www.cfr.org/china/expanding-china-africa-oil-ties/p9557; "China-Africa Trade up 45% in 2008 to US$107 billion", *China Daily* (Beijing), 11 February 2009; International Monetary Fund, *China-Africa Trade Figures* (Washington, DC: IMF, March 2010).
6 Evan S. Medeiros, "Chinese Foreign Policy: The African Dimension", presentation at the FLAD-IPRI conference on "Strategy and Security in Southern Africa" held in Lisbon, Portugal, October 2009, www.ipri.pt.
7 C. Alden, "China in Africa", *Survival*, Vol. 47, No. 3 (2005); P. Mooney, "China's African Safari", *YaleGlobal*, 3 January 2005.
8 Ian Taylor, *China's New Role in Africa* (Boulder, CO: Lynne Rienner, 2009), pp. 1–36.

9 Elizabeth Economy and Michael Oksenberg, *China Joins the World: Progress and Prospects* (New York: Council on Foreign Relations, January 1999), pp. 266–93.
10 Flynt Leverett and Pierre Noel, "The New Axis of Oil", The National Interest (Washington, DC and Sacramento, CA: New American Foundation), www.newamerica.net.
11 John Mitchell, "World Oil Trade–New Oil Axis", *The World Today* (London), March 2010, pp. 9–11; and David Greenaway, *The World Economy: Global Trade Policy 2009* (Hoboken, NJ: Wiley-Blackwell, May 2010).
12 "Rising Angola: Oil, Glorious Oil", *The Economist* (London), 28 January 2010, available at www.economist.com/node/15401935 (accessed 11 July 2012).
13 "Rising Angola: Oil, Glorious Oil", *The Economist* (London), 28 January 2010, available at www.economist.com/node/15401935 (accessed 11 July 2012).
14 V. Brittain, *Death of Dignity: Angola's Civil War* (London: Pluto Press, 1998), p. 85; J. Duffy, *Portuguese Africa* (Cambridge, MA: Harvard Press, 1959); D. Birmingham, *Portugal and Africa* (New York: Palgrave Macmillan, 1999).
15 Sonangol Company History, www.sonangol.co.ao (accessed 11 March 2010).
16 Alex Vines, Lillian Wong, Markus Weimer and Indira Campos, *Thirst for African Oil: Asian National Oil Companies in Nigeria and Angola* (London: Chatham House Report, August 2009), pp. 31–59.
17 *Evolução do mercado petrolífero internacional* (Luanda: Ministério dos petróleos, Angola, 2009), pp. 1–6.
18 Editorial, "Nigeria's Change of Gear", available at www.offshore-technology.com/features/feature81868/feature81868-3.html (accessed 11 July 2012).
19 Ibid.
20 OPEC admitted Angola as its 12th member on 1 January 2007, and in 2009 it took over OPEC's presidency.
21 EIU, "Angola: Country Fact Sheet", Economist Intelligence Unit, *The Economist* (London), 16 February 2010.
22 Paul Collier, *The Bottom Billion: Why the Poorest Countries are Failing and What Can Be Done About It* (New York: Oxford University Press, 2008); David Sogge, "Angola Failed Yet Successful", working paper 81 (Madrid: FRIDE, April 2009), pp. 1–38.
23 Steven M. Gorelick, *Oil Panic and the Global Crisis: Predictions and Myths* (Hoboken, NJ: Wiley-Blackwell, November 2009).
24 *The Economic Report for Africa*, jointly published by the Economic Commission for Africa and the African Union Commission, shows that Africa's 2010 GDP was expected to grow at a rate of 4.3%, www.uneca.org.
25 Paul Stevens, "Prospect for Oil Prices – Up and Down", *The World Today* (London), March 2010, pp. 12–14.
26 Ibid.
27 "New OPEC Chairman Sees Crude Oil at $80 per barrel Reasonable", *Xinhua News* (Beijing), 23 December 2009.
28 OPEC, *Monthly Oil Market Report – February 2010* (Vienna: OPEC Research Division, 10 February 2010), p. 5, www.opec.org/library (accessed 11 March 2010).
29 "Rising Angola: Oil, Glorious Oil", *The Economist* (London), 28 January 2010, available at www.economist.com/node/15401935 (accessed 11 July 2012).
30 "China Plays Predominant Role in Angola's Economic Development", *Angop* (Luanda), 3 July 2009.
31 "Angolan Ambassador Praises Bilateral Cooperation", *Angop* (Luanda), 4 March 2010.
32 For related arguments, see: Irakli Khodeli, *From Curse to Blessing?: Using Natural Resources to Fuel Sustainable Development* (Hoboken, NJ: Wiley-Blackwell, August 2009); and Paul Collier, *The Bottom Billion*, p. 86.
33 Paul Kagame, "Why Africa Welcomes the Chinese", *Guardian* (London), 2 November 2009.

34 "Angola: China's African Foothold", *BBC News* (London), 20 June 2006.
35 David White, "Special Reports: African Infrastructure 2006", *Financial Times* (London), 21 November 2006.
36 Transparency International's Corruption Perception Index 2011 ranked Angola 168 out of 183 countries and territories. See: cpi.transparency.org/cpi2011/results/ (accessed 20 July 2012).
37 Indira Campos and Alex Vines, "Angola and China: A Pragmatic Partnership", Working Paper (Washington: Center for Strategic and International Studies, March 2008), p. 18, www.csis.org.
38 "China's Trade with Portuguese-speaking Countries Doubles in January", *Macauhub* (Macau), 4 March 2010.
39 Editorial, "Angola exporta 29.9 bilioes de dolares em petroleo", *Portugal News*, 8 August 2007.
40 "Angola Began Year as China's Top Oil Supplier", *Macauhub* (Macau), 8 March 2010.
41 Bill Dodson, *China Inside Out: 10 Irreversible Trends Reshaping China and its Relationship with the World* (Hoboken, NJ: Wiley-Blackwell, December 2010).
42 Sigfrido Burgos Cáceres and Sophal Ear, "China's Strategic Interests in Cambodia: Influence and Resources", *Asian Survey*, Vol. 50, No. 3 (May/June 2010); "China World's Biggest Energy User", *The Times* (London), 20 July 2010.
43 National Bureau of Statistics of China, statistical data, www.stats.gov.cn (accessed 11 March 2010).
44 Erica Downs, "China", Energy Security Series at Brookings Foreign Policy Studies (Washington, DC: Brookings Institution, December 2006), pp. 1–67; Taylor, "China's Oil Diplomacy in Africa".
45 Facts and Details, *Chinese Oil and Gas and Natural Gas Imports, Overseas, Soucres and Politics, Chinese oil Imports*, available at factsanddetails.com/china.php?itemid=1727&catid=13&subcatid=85 (accessed 11 July 2012).
46 Bobo Lo, *Axis of Convenience: Moscow, Beijing and the New Geopolitics* (Washington, DC: Brookings Institution, 2008); Matthew Forney, "China's Quest for Oil", *Time* (New York), 18 October 2004, p. 134.
47 US Energy Information Administration, *China – Analysis – Oil* (Washington, DC: EIA, 2012), www.eia.gov/countries/cab.cfm?fips=CH (accessed 13 September 2012); Michael Economides, "China's Oil Imports Continued Upward Climb in 2009", *Energy Tribune*, 26 January 2010.
48 Evan S. Medeiros, *China's International Behavior: Activism, Opportunism and Diversification* (Santa Monica, CA: RAND Corporation, 2009), p. xvii.
49 Lucy Ash, "China in Africa: Developing Ties", *BBC News* (London), 4 December 2007; James Traub, "China's African Adventure", *The New York Times*, 19 November 2006.
50 Erica S. Downs, *China Security: The Fact and Fiction of Sino-African Energy Relations*, July 2007, available at www.cfr.org/china/china-security-fact-fiction-sino-african-energy-relations/p15191 (accessed 11 July 2012).
51 Ibid.
52 David Zweig and Bi Jianhai, "China's Global Hunt for Energy", *Foreign Affairs*, Vol. 84, No. 5 (2005), pp. 26–28; and Jim Rogers, *A Bull in China: Investing Profitable in the World's Greatest Market* (Hoboken, NJ: Wiley-Blackwell, December 2007).
53 Evan S. Medeiros, *China's International Behavior: Activism, Opportunism and Diversification* (Santa Monica, CA: RAND Corporation, 2009).
54 With 75% of GDC shares owned by the China International Fund, it features as the controlling partner.
55 Daniel Balint-Kurti, "Guinea – Bought by Beijing", *The World Today* (London), March 2001, pp. 15–17.

56 Indira Campos and Alex Vines, *Angola and China: A Pragmatic Partnership*, p. 5, available at csis.org/files/media/csis/pubs/080603_campos_angolachina.pdf (accessed 11 July 2012).

57 Lee Levkowitz, Marta McLellan and J. R. Warner, *A Case Study in Chinese Investors' Operations in Angola and Beyond* (Washington, DC: US-China Economic and Security Review Commission, 10 July 2009), p. 52.

58 "Asian Disruptions", *Energy Compass* (New York), 11 December 2009; Vines *et al.*, *Thirst for African Oil*, p. 50.

59 See: Vines *et al.*, *Thirst for African Oil.*

60 "Angola and China are Scoring Points in Africa", *Novo Jornal* (Luanda), 7 November 2009.

61 The Sinopec Group, the largest shareholder of Sinopec Corpn, is a super-large petroleum and petrochemical group incorporated by the state in 1998 based on the former China Petrochemical Corporation. Funded by the state, it is a state-authorized investment arm and state-owned controlling company.

62 Alex Vines, Lillian Wong, Markus Weimer and Indira Campos, *Asian National Oil Companies in Angola: Thirst for African Oil*, p. 33, www.chathamhouse.org.uk (accessed 14 July 2012).

63 "Nigeria: Jonathan Struggles for Power", *Energy Compass* (New York), 5 March 2010.

64 David Barboza's article ("State-Owned Group Fuel China's Real Estate Boom", *The New York Times*, 1 August 2010) notes: "All around the nation, giant state-owned oil, chemical, military, telecom and highway groups are bidding up prices on sprawling plots of land for big real estate projects unrelated to their core businesses … in China, it's the government that controls the money supply and the spending".

65 For broader analysis, see: Thomas Friedman, "The First Law of Petropolitics", *Foreign Policy*, May/June 2006.

66 Wan Zhihong, "Wen Heads 'Super Ministry' for Energy", *China Daily* (Beijing), 28 January 2010.

67 "Clinton Seeks Better Ties with Oil Giant Angola", *China Daily* (Beijing), 10 August 2009.

68 Chevron and Exxon Mobil are already there; Dole Food Co. and Chiquita Brands International want access.

69 "China's Resource Hunger", *The Economist* (London), 13 March 2008.

70 Thomas M. Kane and Lawrence W. Serewicz, "China's Hunger: The Consequences of a Rising Demand for Food and Energy", *Parameters*, (2001), pp. 63–75.

71 Roger Thurow and Scott Kilman, *Enough: Why the World's Poorest Starve in an Age of Plenty* (New York: Public Affairs, 2009), pp. 265–68; M. P. van Dijk, *The New Presence of China in Africa* (Amsterdam: Amsterdam University Press, 2010).

72 Stephanie Kleine-Ahlbrandt and Andrew Small, "China's New Dictatorship Diplomacy: Is Beijing Parting with Pariahs?" *Foreign Affairs*, (January/February 2008), pp. 1–7; Taylor, "China's Oil Diplomacy in Africa".

73 "The Oil Industry, Beyond the Black Stuff, Big Oil is Being Forced to Rethink its Future", *The Economist* (London), 4 February 2010, available at www.economist.com/node/15473681 (accessed 14 July 2012).

74 Ibid.

75 Ian Taylor, *China and Africa: Engagement and Compromise* (London: Routledge, 2006).

76 Janice Jiggins, "Vulnerabilities and Tensions in Global Food Systems: Increasing Resilience or Fragility?", paper presented at the UN Food and Agriculture Organization (FAO), Rome, Italy, 3 March 2010.

77 These notions have been echoed by 24 experts who explored the institutional impacts, dimensions and consequences of the global economic crisis of 2007. See

Anton Hemerijck, "The Institutional Legacy of the Crisis of Global Capitalism", in Anton Hemerijck, Ben Knapen and Ellen van Doorne (eds), *Aftershocks: Economic Crisis and Institutional Choice* (Amsterdam: Amsterdam University Press, 2010), pp. 13–15.

78 J. Stiglitz, A. Sen and J. P. Fitoussi, "Report by the Commission on the Measurement of Economic Performance and Social Progress", Paris-France, September 2009, www.stiglitz-sen-fitoussi.fr (accessed 11 March 2010).

4 Brazil

China is presently the world's foremost energy consumer, with highest growth rate in the global automobile market and is in competition with the USA for global gas, oil and mineral resources, especially in Africa, Central Asia, the Middle East and Latin America.[1] China aims to increase its global access to natural resource reserves, such as iron-ore and minerals in order to supply fast expanding sectors of manufacturing and industry.[2]

For more than a decade, China's economy has grown at around 10% annually based on low-cost labour[3] and the successful penetration of consumption-oriented Western markets.[4] The data portray this growth: the gross domestic product (GDP) growth in the last seven years was 10.1% in 2004, 9.9% in 2005, 11.1% in 2006, 11.4% in 2007, 9.0% in 2008 and 8.7% in 2009, and an estimated 9.5% in 2010.[5] China's strong economic growth has resulted in the country becoming the largest energy consumer. Between 2000 and 2010, non-OECD (Organisation for Economic Co-operation and Development) oil consumption increased more than 40%, with China, India and Saudi Arabia accounting for the largest growth in oil consumption among the countries in the non-OECD grouping during this period. A rise in China's consumption is forecast to grow from 5 million to 13 million barrels per day in 2015. Imported oil will become a necessity as domestic sources are not commensurate with growth in demand.[6] In view of this evolving situation, Beijing tries to outwit big international resource companies by aggressively seeking out potential deals in countries that, for the most part, would benefit greatly from a steady and long-term demand for their commodity exports. In such a situation, the Chinese Government aims at avoiding global resource businesses in favour of companies that would benefit from China's exportation of its own commodities over the long term.[7] China helps its companies focused on resources with grants and favourable loans in order to procure contracts.[8] However, companies do not base their aims solely on commercial objectives, but rather they employ combined objectives of marketing and strategy.[9]

Latin American countries, in recognition of Chinese needs and of Beijing's explicit interests in the region, are exploring cooperative arrangements with China in the fields of agriculture, energy, infrastructure, manufacture, mining, and science and technology.[10] Specifically, within Latin America, Brazil is playing not only a regionally but also a more globally assertive role, and this demeanour, some observers say, is nothing new.[11] Throughout the first decade of the 21st century, Brazil has continued to consolidate its position as the most influential actor in South America without raising any sort of hegemonic suspicions. This assertiveness and influence is partly explained by Brazil's young, vibrant, multicultural and reasonably educated population,[12] which offers perennial promises for economic development and social progress, as does the country's vast deposits of oil, gas and minerals, forests and soybean plantations. The foreign perception is that Brazil is emerging as the "country of the future", a nation-state that, if it continues on the steady improvement track, is poised to replace Germany as the world's fifth largest economy by 2016.[13] To Brazil, at this moment, successfully managing the ongoing economic boom is a priority, as in the past it has unintentionally squandered its vast potential to grow and reach great-power status.

In general, South America has a history of producing and exporting oil for more than a century, with most of its shipments flowing north, to the USA; but as globalization emerged in the 1980s, shipments started to find other destinations. As international energy markets homogenized the range of players buying and selling resources, countries like Brazil were able to place their oil at the disposal of China, or India, or any other country willing to buy oil.[14] It is for this reason that in recent years, China's voracious economy has brought it to Brazil, and much of South America, in search of energy sources and vast natural resources.

China, though, is not just interested in Brazil.[15] To Beijing there is no such thing as reconciliation exclusivity. Much of South America has become crucial to China's need for additional natural resources, raw materials and more markets for its products.[16] For example, it is interested in copper, nickel and tin from Chile, natural gas from Bolivia, oil from Ecuador and Venezuela, and precious woods from Colombia. As a sub-sector, mining is among China's top priorities. So relevant is China's nascent presence in the region that in response to Beijing's direct requests, "Argentina, Brazil, Chile, Peru and Venezuela have conferred on China the status of 'free-market economy' thereby strategically decreasing the potential impact that anti-dumping measures may have on inexpensive Chinese imports".[17]

To give an idea of China's strategic interests in the region, in 2008 Chinese President Hu Jintao visited South America as part of a country-hopping travel schedule with a dozen government ministers and hundreds of businessmen and industry representatives to offer, negotiate and sign a total of 39 trade agreements and memorandums of understanding related to energy, minerals and agricultural commodities. This tactical attention is not at all surprising: China's annual regional trade is over $140 billion,[18] and because Beijing's economic priorities have nothing to do with the promotion of democracy and human rights, heads of South American states are lured in by this "no strings attached" approach.[19] Moreover, "China's foreign direct investment in Latin America reached roughly $25 billion in 2008–9, making up 14.6 percent of China's total foreign direct investments".[20]

Brief outline of this chapter

This chapter commences with a list of Brazil-China treaties, pacts and accords in order to give a better understanding of the pair's mutual interests. Following this, we pragmatically analyse several devices and instruments that enabled achieving our aims and targets. In concluding, we give summarized results and attempt to illuminate contributions provided by our work in comprehending Latin American interstate relations regarding scholastic studies.

The Brazil-China connection: results following actions

Importantly, Brazil has replaced the USA with China as the largest market for exportations and partner for trade, including commodities such as beef, chicken, coffee, orange juice, sugar and tobacco of which Brazil is the global leader in exports and their strategic concomitancy includes energy security. China has foreign reserves in excess of $3.3 trillion and Brazil feels the benefits of China's external strategy of scouring Brazil's territories in an opportunistic manner.[21] In addition to natural gas and oil exploration, Chinese investments involve a factory to manufacture oil and gas equipment, an automobile plant, a steel mill, a shipyard and port infrastructure development. In Beijing, government officials push aggressively to invest in foreign manufacturing and the industrial sector to bolster the country's presence, its image and its political influence.[22]

To gain some perspective on Sino-Brazilian bilateral relations, Kerry Dumbaugh and Mark P. Sullivan note that the "China-Brazil picture differs completely from that of merely seven years ago. In 2004, China's imports from and exports to Brazil amounted to US$8.7 billion and US$3.7 billion, respectively. But even then, Beijing was already making announcements of tens of billions of dollars in energy-related deals and investments".[23] In order to help understand to what extent China has invested in Brazil, "in the first eight months of 2010, Beijing invested over $20 billion in Brazil", 10 times more than all of China's previous investment in the country.[24] Chinese technocrats and industrialists have developed a deep understanding of Brazilian political economy. They have quickly learned the strong links between people, money, respect and political power. Armed with this knowledge and with plenty of cash to spend, many Chinese state and non-state companies have arrived in Brazil to negotiate a modern high-speed rail construction contract, to buy huge tracts of land in order to plant soybeans and other crops, to build modern vehicle motor manufacturing plants, to realize telecommunications infrastructure, to construct high-volume steel mills and to purchase a stake in Brazil's electrical grid system. If this continued, China was on track to become Brazil's number one investor for 2010 and 2011.[25]

The following examples demonstrate some of the Chinese transactions in Brazil that so far have been reported by national and international sources:

> The East China Mineral Exploration and Development Bureau, a quasi-governmental mining entity backed by the Jiangsu province in China, signed in early March 2010 an agreement with Itaminas Comercio de Minérios SA to acquire the company along with its iron-ore mining operations in the Brazilian eastern state of Minas Gerais for $1.2 billion. This company buyout was driven by the Itaminas mine which has an estimated 1.3 billion tons of iron-ore reserves,[26] with mineral outputs flowing via direct railway to Rio de Janeiro's port.[27]

Also, in Porto do Acu, 175 miles north of Rio de Janeiro, port "workers are busy building a two-mile long pier to accommodate"[28] the mega-tankers to transport crude oil and the monumental vessels that will transport iron-ore and other minerals to China's ravenous metallurgic industries. However, the São Paulo State Industrial Federation claims "unfairness" and has called on the Brazilian government to take serious steps to stop the wave of "purchases of local mining assets by Chinese companies".[29] There are no surprises here, because "it is well known in foreign investment circles that Chinese state-owned companies have an edge over American and other multinational energy companies bound by internationally-mandated restrictions related to anti-corruption, accountability, bidding and transparency that are intended to promote fair competition. It seems that, when it comes down to overseas investments, China plays by its own rules".[30] This is a fact that local industrialists are holding onto in order to press their claims.

We now move away from mining in order to discuss the oil industry. China Petroleum and Chemical Corporation (Sinopec) in Beijing is among the most prominent of Chinese firms involved in crude oil exploration, production and distribution on a global scale. "Sinopec controls dozens of subsidiaries around the world, including the US-listed Sinopec Corp., with principal businesses in exploration, production, and trading of petroleum and natural gas, refining and sales of petroleum products, and sales of chemicals".[31] With production levels of 296.8 million barrels of crude oil and 293.06 billion cubic feet of natural gas in 2008, Sinopec is the second largest producer of crude oil and natural gas in China.[32] The other behemoth is the China National Offshore Oil Corporation (CNOOC).[33] These two corporations are active and operational in Brazil, seeking and securing as many natural resource deals as possible. More precisely, their presence in Brazilian territories is justified by the developing trillion-dollar story around oil that is estimated to raise Brazil's production levels to over 6 million barrels per day. At this level of output Brazil could easily become the third or fourth producer of crude oil in the world and a massive exporter. We will now inspect and discuss these dynamics.

As mentioned previously, partly, Sinopec and CNOOC exist because Brazil's most substantial crude oil reserves lie 150 miles offshore and a few miles beneath ocean, rock and unstable layers of salts and geological formations. Reaching it requires the creativity and expertise of the oil industry's top companies.[34] Brazil's national oil company, Petroleo Brasileiro SA, or Petrobras, recently discovered a colossal reserve[35] of crude oil buried 7,000 metres "beneath a thick layer of salt on the Atlantic Ocean floor",[36] which sparked an internecine strife among Brazilian provinces competing for their fair share of royalties.[37] Petrobras—Brazil's main oil producer—along with Sinopec and CNOOC[38] have studied the feasibility and viability of joint operations in exploration, processing and distribution of this and other oil reserves in their respective countries, and also in other regions of the world too.[39] Furthermore, in the first quarter of 2010, the Brazilian Government reached a binding agreement on the order of $1 billion with Sinopec to build a natural gas pipeline that will cross Brazil. "In addition to this agreement, Chinese officials are intently listening to rumours of plans to build a cross-country pipeline to transport crude oil to the Pacific where it would be loaded onto mega-tankers that are too big to use the Panama Canal. Another speculative proposal involves Colombia",[40] laying a pipeline to carry hydrocarbons to Colombia's Pacific ports for shipments to Asia.[41] "All these ideas seem favourable under a climate of high gas and crude oil prices. For its part, Beijing prefers to deal directly with governments in oil-producing countries and to use its vast wealth to reduce the role of big foreign oil companies that have long been the traditional intermediaries (and price inflators) between oil producers and oil consumers".[42]

To consolidate further bilateral relations and future investments, in April 2010 Brazil and China signed a joint action plan for 2010–14 and reached agreements in several economic sectors.[43] However, foreign investments and bilateral trade are not always smooth. For instance, in Brazil, manufacturers and industrialists have accused Beijing of dumping products in their markets and this has caused an avalanche of calls to the Brazilian Government to levy new tariffs on some Chinese imports.[44] Also, several neighbouring countries[45] have voiced their concerns about "Beijing's conspicuously aggressive efforts to win vast access to energy reserves and mineral deposits in the region".[46] More specifically, China's entry to Brazil (and Venezuela) raises anxiety among South American countries because of the fear of becoming less attractive markets to local oil-producers and, thus, more dependent on foreign oil imports at a time when international reserves remain relatively tight. On a more individual level, energy experts suspect that Chinese engineers and resource officials will use accords, deals and pacts as a tool to learn as much as possible from Brazilian and Venezuelan heavy-oil refining technologies that for years have been adapted to local conditions: access to confidential operational information grants China more leverage.

In addition to an accord with Brazil, Beijing has signed investment deals with Venezuela and Peru,[47] "both of which represent largely untapped markets, and is exploring possible arrangements with Bolivia and Colombia".[48] "One example can illustrate that Brazil is not unique in sealing multibillion dollar deals: in early 2010 the CNOOC, eager to expand its crude oil reserves to meet increasing domestic demand, paid $3 billion for a 50 percent stake in an Argentinean privately-owned oil and gas group called Bridas".[49]

In summary, "China is driven towards Brazil by its need for privileged access to raw materials, minerals (iron-ore), and hydrocarbons. Obviously, Brazil and China are both emerging market economies that share similar challenges in international markets, and these similarities inexorably link the two countries' interests. But still, some hesitation lingers. After studying the actions, behaviours, and consequences of some overseas investments by the China National Petroleum Co. and Sinopec",[50] Brazil and others are quickly "realizing that Chinese companies are competing on their own turf for natural resources and market shares".[51]

Economic diplomacy in a changed world: so far, so good

We noted in the last section, that China has enjoyed global resources and is attracted to Latin America, and particularly Brazil, for energy and other resources, counterbalancing the USA's lack of interest in Latin America. "This benign neglect is all the more disconcerting if one looks at the region as a potential market block that has increased its purchasing power: from 2003 to 2008 Latin America enjoyed a prolonged period of healthy growth averaging 4.3 percent with low inflation and a balance-of-payments surplus".[52] In the second decade of this still young millennium, Latin America is positioned to rise as a vibrant international market and as a suitable destination for foreign investments.

Regionally, as well as globally, the balance of power is starting to shift to countries like Brazil, the political clout and economic weight of which has grown considerably in recent years.[53] For its part, the European Union (EU) has taken note of these commercial and diplomatic overtures. This is not surprising given that Europe is still Latin America's biggest foreign investor and aid donor; however, Chinese trade with the region has grown much faster than Europe's over the past decade.[54] Furthermore, Brasilia has been clever in raising its diplomatic game, acting the part of an influential regional power with a significant global impact. This comes in part because it is the "B" in the vibrant economic club of Brazil, India, Russia and China (BRIC), thus garnering the country leverage and high visibility. Furthermore, it has successfully integrated into the global economy by boosting exports, stabilizing its currency and polishing its international image.[55]

During the Cold War, countries like Brazil—as well as China and India—tiptoed and carefully aligned themselves according to the prevailing balance of power in the bipolar competition between the Soviet Union and the USA. This is no longer the case. New patterns and habits of international coordination and cooperation are emerging in place of deference to US leadership. China is becoming a partner of choice. Another reason for preferring China over the USA is that relations between Brazil and the USA have lost momentum as a result of growing differences over policies and trade imbalances. These differences threaten to upset the delicate balance with a country that is pivotal to American interests in its closest region.[56] Also, several instances of strategic marginalization are taking place. The rationale behind this is that, by excluding the USA, smaller groups of Latin American countries are learning to operate within deepening patterns of regional economic cooperation and political coordination. These groups of countries are either seeking stronger cohesion among them or welcoming economic and diplomatic overtures from China (also from India and Russia). Commentators believe that in adopting a low-key approach and tactically managing relations, confrontations within the Western Hemisphere can be largely avoided.[57] Some nations around the world have also demonstrated their explicit preference for Chinese investments by acceding to Beijing's whims and desires,[58] while others have been more conservative while remaining active in bilateral trade. One of the reasons for this sustained bilateral trade is that China's official trade barriers are lower than those in other big emerging economies. For example, in 2008 the average applied tariff was 9.6% in China compared with 13% in India, and 13.6% in Brazil. On the other hand, according to estimates from the UN Conference on Trade and Development (UNCTAD), non-tariff barriers such as import quotas or restrictions on foreign involvement in government tenders were equivalent to an added tax of 9.4% on imports into China.[59]

So, one may ask, how is it possible that China is received with open arms and able to invest wherever it goes? The answer to this question lies in straightforward arithmetic. Given that financing can often account for two-fifths of the cost of an entire project, exceedingly low-interest rate credit lines are attractive to any foreign investor, regardless of geographical location. Knowing this key fact, China expertly uses low-interest rate financing. It furnishes credit accords and loans in the order of billions of dollars at incredibly low interest rates that can stretch beyond 20 years.[60] The size of Chinese banks is so large that such banks have the financial capacity not only administratively to handle these loans but also to support a major investment expansion of Chinese firms overseas. All this underscores China's importance as an emerging lender of last resort rivalling the International Monetary Fund (IMF). At a disadvantage, however, are the member countries of the OECD, which agreed long ago not to use low rate financing as a competitive tool to secure deals.[61]

The confluence of global financial markets on the verge of collapse and a tight credit environment has enabled China to seal enviable deals and secure its future. Evidently, in the usual pragmatic utilitarianism that often characterizes emerging powers, Brazil and China play down areas of bilateral disagreements, and focus on shared commitments to economic stability, trade promotion and energy security—two countries with long roads ahead of them.

A key component of China's strategy to guarantee access to Brazilian resources is to build close ties with top officials via high-level visits of Chinese leaders leading to pronouncements of strategic partnerships between countries. Moreover, a government change in Brasilia could lead to a reduced reliance on the USA as sole guarantor of Brazil's security and offer China an expanded role. Western analysts believe that China's continued penetration into the western hemisphere could have profound economic and political implications for the USA. With less oil available to the American market, the USA could be forced to start seeking oil elsewhere, primarily in the Middle East, hence becoming more dependent on an already belligerently volatile region.[62]

The Jintao administration, with its rapid growth and widespread tentacles, now has the opportunity to serve as an "enabler" of initiatives for deeper cooperation beyond energy in a number of regions of the world where this is still weak, from Africa to Latin America to Southeast Asia.[63] At the same time, the Brazilian Government can use its nascent leadership and diplomatic arsenal to tackle global climate change which is borderless and involves a multiplicity of private and public stakeholders. Furthermore, Brazil is eyeing improvements in its public sector to climb closer to the levels attained in rich, industrialized nations. The coherent utilization of gas, minerals and oil is assisting in fulfilling those dreams given that high commodity prices can strengthen government finances and allow for a big increase in social spending. The Lula Government's platform to win the elections back in 2002 focused much on social programmes[64] that reached out to the poor and rural populations at the margin of society,[65] and these projects and programmes are positioned to be strengthened,[66] in light of the spill-over effects of China's incoming economic and energy diplomacy.

In summary, there are numerous countries with which China can establish commercial relationships, but not all of them carry tactical relevance or strategic interest. As a whole, China adapts the style and substance of its energy and foreign policy to the emergence of a more interconnected and multipolar world order.[67] Beijing uses its huge foreign reserves and low-interest rate loans to muscle its way through Brazil and other South American countries. Big economies in Latin America are aware of the opportunities that a long-lasting relationship with China can bring and most of them are snatching them on the spot. China's capital investments mean job creation, increased trade and sustained economic growth. In a way, Brazil follows a path similar to its neighbours but, at the same, differs from them. The Government of Lula da Silva seeks more than just economic benefits. Now that it is part of BRIC, the elite group of emerging market economies, it has much bigger ambitions. Brazil seeks a more mature, "steady and solid relationship with Beijing", based on a mutually beneficial partnership, instead of the last five decades of imposed leadership from Washington.[68]

The deepening cooperation between Brazil and China also serves as a further catalyst to the process of growing transcontinental consultation on addressing common global challenges. These challenges are related to climate change, the collective management of the global economy,[69] non-proliferation of nuclear weapons,[70] and the UN within an international legal framework. A careful allocation of attention between novel and old powers (i.e. China and the USA) along with judicious distribution of economic diplomacy may very well guide the necessary policy adjustments that might offer Brazil the best chance to play a positive role in the areas meaningful for the Lula administration.

Conclusion

China has experienced and continues to experience tremendous economic growth. To sustain this growth, it must keep the wheels of its manufacturing and industrial sectors turning. As a result, it is "driven by the need for privileged access to raw materials and hydrocarbons".[71] In this global search it has found Brazil, a country with vast lands and natural resources, but it turns out that Brazil has found China too, a fellow BRIC.

Obviously, for these countries to establish a cordial working relationship they must want something from each other. In general, these interests can be diplomatic, economic, political, military, social, physical, strategic and/or technological, among many others. From their actions and policies one can try to comprehend their incentives and motivations. For example, the Sino-Brazilian connection appears to be grounded in diplomatic, economic, physical and strategic interests. Below follows a brief examination of these specific interests.

Diplomatically, Beijing is bolstering China's diplomatic presence in the Southern cone as a manoeuvre to limit—or at least make difficult—Taiwan's relations[72] with Latin American countries with which it has active relations.[73] A reunified China is a paramount ambition.[74] For its part, Brazil feels no compunction to verbalize its global ambitions. One of these revolves around a permanent seat on the UN Security Council.[75] To make public display of its solidified credentials, Brazil views the discourse of anthropogenic global warming as an opportunity to exploit its comparative advantages as premier exporter of sugarcane-based ethanol and, arguably, a defender of the Amazon rainforest. Most passionately, like China, Brazil wants to be regarded and treated as an equally important, responsible and consultable player on the international stage.[76]

Economically, both Brazil and China are strategically and structurally well positioned to emerge vibrant and strong from the current economic slowdown[77] and, as a consequence, to continue to underpin their long-term potential for shared global leadership. As of this writing, "China is Brazil's largest trading partner and biggest export market".[78] Furthermore, by mid-2011, "China will displace the European Union as Latin America's second-largest trading partner".[79] Moreover, what has been seen in the last decade is that the Chinese economic model is based on heavy importation of commodities, including agricultural items, and using those commodities very efficiently in manufacturing and industrial processes to produce goods. These goods—which were produced by paying low wages—are then exported to rich countries. This has proved a very lucrative business for the Chinese. However, for the system to work smoothly into the future, it must have unrestricted access to commodities. In this context, disruptions to raw material supplies are viewed as vulnerabilities, so part of the reason for Beijing's *convenient marriage* of economic, foreign and energy policies and strategies is to buy into the supply channels of critical commodities all around the world.

Physically, China's global resource quest emphasizes securing energy sources (natural gas and crude oil), natural resources (agricultural commodities, rubber and timber), and raw materials (minerals such as copper, gold, iron-ore, nickel, silver and tin). In moving ahead with plans for resource-related investments, the Chinese sit at South American negotiation tables without human rights or good governance demands. They arrive, ready to invest billions of dollars and to negotiate the best possible terms of transaction. In Brazil, Beijing has landed accords and deals on natural gas and crude oil explorations, railways, soybean farming, manufacturing plants, telecommunications infrastructure, steel mills and electricity generation projects. International commentators and scholars critical of China's global resources quest fail to realize that Chinese funding of gas and oil projects around the globe ultimately means cheap energy for the global commons. This is because most of the world's natural gas and crude oil are sold on the international spot markets to the highest available bidder and China's largesse of extending preferential credit lines to gas- and oil-producing countries should keep gas and oil prices from increasing simply by the mere fact that global supply of gas and oil will rise (especially once Brazil starts pumping 6 million barrels daily).

Strategically, Beijing is allocating monies cautiously in a broad range of diversified investments, with larger ambitions in mind. As per Chinese work style, these are well thought out steps that are prerequisites for a longer-term relationship, yet at the same time they are small enough (in comparison to other projects elsewhere, e.g. in Nigeria) easily to be called up should the economic, political, or social conditions dramatically change in a region that has not had a history of deep reconciliation with East Asia.[80] Geopolitical interests also fit under the auspices of strategic interests. China, being communist, identifies easily with leftist approaches to governance. In relation to this, commentators note that in essence there are two left wings in Latin America. The first is a moderate left once led by Lula da Silva.[81] The second is a radical left led by Hugo Chávez in Venezuela.[82] Most importantly, Brazil and Venezuela are not hostile to each other. Furthermore, at Brazil's insistence, Venezuela was invited to join Mercado Común del Sur—Common Market of the South (MERCOSUR). The two countries are the driving force behind a plan for regional defence, a project for political integration and a new development bank for South America. The ability of Chávez and Lula da Silva to work together across a broad range of interests signals South American leaders' desire to reduce US influence and to enhance the latent capacity of the sub-region to diligently resolve its own problems without disruptive interference from external sources.[83] Although geographically speaking the countries south of the Rio Bravo (part of the Mexico–US border) are still in the US sphere of influence, by itself geography no longer determines destiny. This is most vividly evidenced by China's reconciliation with Brazil, and other Latin American countries, which acted assertively on opportunities to build bridges in areas where US influence had long been on the decline.[84]

From afar, it all looks and sounds good, positive and rosy, but a closer examination of information sources reveals that there are a few kinks in the narratives that have surfaced in response to Chinese presence in the region related to dumping of products, loss of market share, unfair competition and corruption allegations. Also, critics of this newfound strategic union are finding ways to spread the notion that China is more of a strategic competitor than a strategic partner. Luckily for critics, this sentiment is gaining traction through region-wide access to radio, online blogging, newspapers and a growing range of satellite television channels that help shape opinions and reactions to Chinese investments in Brazil and South America. The USA is also raising sensitive issues. In addition to concerns over energy security, American analysts and US officials have expressed anxiety regarding Brazilian-Chinese cooperation in satellite and rocket technology.[85] Moreover, the West, in general, has grown tired of China's currency value suppression—that is, the undervaluation of the yuan—and repeatedly has asked China to revalue its currency, to little avail.[86]

The narrowing of the relative levels of economic, military and political power between America and the rest of the world has created a more levelled arena for Brazil and China to reassert their newfound confidence and international influence. At least, in the near future, there is little likelihood that Sino-Brazilian relations are going to cease. There is hope, however, that this alliance will be beneficial to the world at large. In the end, this chapter places the Brazil-China connection in terms of expediency, pragmatism and utilitarianism that reflect the mutually beneficial outcomes of reconciliation. The body of scholarly work in Latin American studies benefits from clearer pictures in non-traditional interstate intercourse.

Acknowledgement

This chapter is adapted from a refereed article entitled "China's Natural Resource Appetite in Brazil", by the authors, and was published in the *Asian Journal of Latin American Studies*, Vol. 24, No. 2 (2011), pp. 71–92.

Notes

1 David Zweig and Bi Jianhai, "China's Global Hunt for Energy", *Foreign Affairs*, Vol. 84, No. 5 (September/October 2005); "China World's Biggest Energy User", *The Times* (London), 20 July 2010; Heinrich Kreft, "China's Quest for Energy", *Hoover Policy Review*, Issue 139 (October–November 2006), www.hoover.org/publications. While it is true that China is conspicuously implementing a "go abroad" policy, other emerging market economies are in equal need of vast natural resources and are seeking them out.

2 Erica Strecker Downs, *China's Quest for Energy Security* (Santa Monica, CA: RAND Corporation, 2006), p. 17. It is worth mentioning here that Brazil claims to have 8,900 metric tonnes of iron-ore reserves and to produce 250m tonnes per year. In fact, availability of iron-ore can be cited as a powerful driver of Chinese interests in Brazil.

3 Jeffrey Sachs, "The Earth Provides Enough to Meet Everyone's Needs", *The National*, 2 March 2011, available at www.thenational.ae/thenationalconversation/comment/the-earth-provides-enough-to-meet-everyones-needs (accessed 11 July 2012).

4 Barry Naughton, *The Chinese Economy: Transitions and Growth* (Boston, MA: MIT Press, 2007), p. 11; Thomas L. Friedman, *Hot, Flat and Crowded: Why we Need a Green Revolution and How it can Renew America* (New York: Farrar, Straus and Giroux, 2008), pp. 343–66. For more data, see: "China", www.state.gov (accessed 23 December 2010).

5 National Bureau of Statistics of China, www.stats.gov.cn/english/ (accessed 23 December 2010); "China's GDP Grows by Seven-year Low of 9% in 2008", *Xinhua* (Beijing), 22 January 2009; "China GDP Grows by 8.7 percent in 2009", *CNN* (Atlanta), 21 January 2010; Qiang Xiaoji, "World Bank Sees China GDP up 9.5% in 2010", *China Daily* (Beijing), 17 March 2010; "Watching China Whizz By", *The Economist*, 19 August 2010. These growth rates are impressive, but there is a caveat: the double-digit annual GDP gains in the past were relatively easy because the economy was growing from a small base. Those kinds of huge gains will be harder to achieve now that China is the world's second-largest economy, about one-third the size of the USA.

6 US Energy Information Administration, *Energy & Financial Markets: What Drives Crude Oil Prices?* (Washington, DC: EIA, 2012), www.eia.gov/finance/markets/demand-nonoecd.cfm (accessed 13 September 2012). For more information, see: US Department of Energy, www.energy.gov (accessed on 23 December 2010).

7 Although all countries generate significant revenues from commodity export not all countries benefit equally. For instance, in some African countries resource rents are used to fuel rebellions, genocides and armed conflict.

8 Juan Forero, "China's Oil Diplomacy in Latin America", *The New York Times*, 1 March 2005.

9 Despite criticism levelled against China, some commentators argue that China's expansion in Latin America is benign, complacent and moderated, mainly seeking mutually beneficial trade and investment opportunities. See: Guillermo R. Delamer, Lyle J. Goldstein, Jorge Eduardo Malena and Gabriela E. Porn, "Chinese Interests in Latin America", in Paul D. Taylor (ed.), *Latin American Security Challenges* (Naval War College, 2004).

10 According to the most recent report by the Economic Commission for Latin America and the Caribbean (ECLAC), by the middle of 2011, China will displace the European Union as Latin America's second-largest trading partner. In early 2011, China had replaced the USA as Brazil's leading trade partner.

11 John Williamson, "Lula's Brazil", *Foreign Affairs*, Vol. 82, No. 105 (January/February 2003), pp. 105–7.

12 Thomas E. Skidmore, "Bi-racial USA vs. Multi-racial Brazil: Is the Contrast Still Valid?" *Journal of Latin American Studies*, Vol. 25, No. 2 (February 2009), pp. 373–86.

13 President Lula da Silva certainly believes this is possible. He travels around the country inaugurating projects and giving talks at forums telling people that this is indeed so. See: "Brazil to Be the Fifth Largest Economy by 2016—President Says", *Agence France-Presse*, 29 December 2010.

14 House International Relations Subcommittee on the Western Hemisphere, *China's Influence in the Western Hemisphere*, 109th Congress, First Session, 6 April 2005. Note that South and Central America make up Latin America.

15 According to Inter-American Development Bank officials, China's foreign direct investment in the region has so far concentrated in Argentina, Bolivia, Brazil, Chile, Ecuador, Mexico, Peru and Venezuela.

16 Consumers number "560 million and there is a combined GDP of US$4 trillion in this region". See: Sigfrido Burgos Cáceres and Sophal Ear, "China's Natural Resource Appetite in Brazil", *Asian Journal of Latin American Studies*, Vol. 24, No. 2 (2011), pp. 71–92, p. 73.

17 Adam Thomson, "Granting of Market Economy Status to China Opens Argentina's Door to Investments", *Financial Times* (London), 18 November 2004; BBC News, "Chinese Vice President's Visit Promote Ties with Latin America", *BBC Monitoring International Reports* (London), 5 February 2005. For an interesting review of protectionist policies, see: John Coatsworth and Jeffrey Williamson, "Always Protectionist? Latin American Tariffs from Independence to Great Depression", *Journal of Latin American Studies*, Vol. 36, No. 2 (May 2004), pp. 205–32.

18 Between 2000 and 2007, China's trade with Latin America grew 10-fold and reached $142 billion in 2008. This amount is 4.7 times less than America's trade with the region, which in 2008 reached $667 billion. See: P. Brookes, "Russia, China, Move in on Latin America", *Institutional Report*, Washington, DC: The Heritage Foundation, 2008.

19 Peter Brookes, "Russia, China, Move in on Latin America", Washington, DC: The Heritage Foundation, 27 November 2008, available at www.heritage.org/Research/Commentary/2008/11/Russia-China-Move-In-on-Latin-America (accessed 3 September 2010); Also, Forero, "China's Oil Diplomacy in Latin America".

20 Chinese Ministry of Commerce, english.mofcom.gov.cn (accessed 13 November 2010).

21 "According to Brazil's Central Bank, Chinese direct investment in Brazil increased to US$367 million in the first half of 2010 from US$73 million in the same period of 2009. This represents over 500% rise in investments. The Chinese used the global recession of 2008–10 to acquire good companies inexpensively". See: "Brazil Destines $367 Million for Research and Innovation", *NoticiasFinancieras* (Miami), 20 February 2006, p. 1.

22 "It is noteworthy that, in 2010, Chinese companies from Sinochem Group to State Grid Corporation of China announced deals or possible acquisitions in Latin American agriculture, natural gas, mining, petroleum, and electrical power as the nation seeks to secure commodity supplies to feed its rapidly expanding economy", RABOBANK, "Chinese Investments in South American Agribusiness", Industry Note 276-2011.

23 Kerry Dumbaugh and Mark P. Sullivan, "China's Growing Interest in Latin America", Report 22119 (Washington, DC: Congressional Research Service, 20 April 2005), pp. 2–3, available at www.usembassy.it/pdf/other/RS22119.pdf (accessed 23 December 2010).

24 Some of it by China Investment Corporation, www.china-inv.cn/cicen/ (accessed 23 December 2010). J. Pomfret, "China Invests Heavily in Brazil, Elsewhere in Pursuit of Political Heft", *Washington Post*, 26 July 2010.

25 John Lyons, "Brazil Turns to China to Help Finance Oil Projects", *The Wall Street Journal* (Asia), 18 May 2009.

26 Brazil is a leading producer of iron-ore for regional and international markets. In fact, Brazilian mining giant Vale Minerals China Co., a competitor of Itaminas mines, expects to produce 450 million tonnes of very high-quality iron-ore annually by 2015, see: "Vale Sees Iron-ore Output up 50% by 2015", *Reuters*, 18 January 2011.

27 S. Mayner, "Chinese Firm Agrees to Buy Brazilian Iron-ore Mine", *The Wall Street Journal*, 27–28 March 2010.

28 John Pomfret, "China Invests Heavily in Brazil, Elsewhere in Pursuit of Political Heft", *The Washington Post*, 26 July 2010, available at www.washingtonpost.com/wp-dyn/content/article/2010/07/25/AR2010072502979.html (accessed 14 July 2012).

29 "Brazil CSN CEO Calls for China Mining Investment Restrictions", *Dow Jones Newswires*, 22 August 2010.

30 For more specific details, see: John Pomfret, "China Invests Heavily in Brazil, Elsewhere in Pursuit of Political Heft", *The Washington Post*, 26 July 2010; Lyons, "Brazil Turns to China to Help Finance Oil Projects".

31 Sembcorp Marine,1st Jack-up Drilling Rig Order from China's Sinopec International (Hong Kong) Co for PPL Shipyard. Stock Exchange Announcements, 17 September 2008, available at www.sembcorpmarine.com.sg/index.php?mact=News,cntnt01,detail,0& cntnt01articleid=223&cntnt01category_id=111&cntnt01returnid=78 (accessed 14 July 2012).

32 About Sinopec: Our Company, english.sinopec.com/about_sinopec/ (accessed 3 September 2010).

33 CNOOC Ltd, www.cnoocltd.com/encnoocltd/default.shtml (accessed 3 September 2010).

34 Exxon Mobil Corp. and Royal Dutch Shell PLC are the technology-savvy industry leaders.

35 "Arguably the largest discovery of oil reserves in the past 35 years, according to preliminary estimates". For more info, see: "Oil Policy in Brazil: Raining on Rio's Parade", *The Economist*, 8 April 2010, available at www.economist.com/node/15868357 (accessed 14 July 2012).

36 "Oil Policy in Brazil, Raining on Rio's Parade", *The Economist*, 8 April 2010.

37 Ibid.

38 "CNOOC Interest in Brazil Assets Driven by China Need for Energy, Chemicals", *Bloomberg News*, 15 September 2010. To give a dimension of China's recent investments, both CNOOC and Sinopec offered at least $7 billion for Brazilian oil assets and a stake in OGX Petroleo & Gas Participacoes SA of Rio de Janeiro.

39 "Aside from direct investments by China Petroleum and Chemical Corporation and China National Offshore Oil Corporation, Petrobras received US$10 billion from the China Development Bank as a down payment on future business: both the private and public sectors are active in Brazil's investment landscapes". See: Editorial, "CNOOC Interest in Brazil Assets Driven by China Need for Energy, Chemicals", *Bloomberg News*, 15 September 2010.

40 Margaret Ryan, "No More Slow Boat to China for US Liquefied Natural Gas", *AOL Energy*, 27 January 2012, available at energy.aol.com/2012/01/27/no-more-slow-boat-to-china-for-us-liquefied-natural-gas/ (accessed 14 July 2012).

41 Robert Slater, *Seizing Power: The Grab for Global Oil Wealth* (Hoboken, NJ: John Wiley & Sons 2010); Lyons, "Brazil Turns to China to Help Finance Oil Projects"; Pomfret, "China Invests Heavily in Brazil".

42 These are only ideas and potential projects noted by Slater, *Seizing Power*; Lyons, "Brazil Turns to China to Help Finance Oil Projects"; Pomfret, "China Invests Heavily in Brazil".

43 Niu Haibin, *Emerging Global Partnership: Brazil and China*, Shanghai Institute for International Studies, 2010, available at www.siis.org.cn/en/zhuanti_view_en.aspx?id=10099 (accessed 14 July 2012).

44 This is particularly worrisome to Brazilian businessmen as they intend to make the most of a non-commodity boom precipitated by free and open access to old and new markets coupled with a sustained global economic recovery. It must not be forgotten that Brazil has both a solid manufacturing base and a large auto industry. In fact, aviation giant Embraer, the world's third largest aircraft manufacturer, has sizeable operations in Brazil.

45 "Argentina, Chile, Peru and Colombia are wary of the breadth and depth of Chinese investments in their region. They worry that China's rising presence could lead to economic manacling and neo-colonialism". See: M. Moxley, "Beijing Strengthens Links with Latin America", *Asia Times Online*, 20 August 2010.

46 Mitch Moxley, "Beijing Strengthens Links with Latin America".

47 Daniel Erikson, "A Dragon in the Andes: China, Venezuela, and US Energy Security", *Military Review* (July/August 2006), pp. 83–89; Simon Romero, "Chavez Says China to Lend Venezuela $20 billion", *The New York Times*, 18 April 2010. From a geopolitical point of view, it is relevant to note that China supports Venezuela's military and Cuba's air defence systems, two meretricious actions that the West is following closely.

48 Thomas Lum, *China's Assistance and Government Sponsored Investments Activities in Africa, Latin America, and Southeast Asia* (Washington, DC: Congressional Research Service, November 2009), pp. 12–14.

49 See the Business section of "The World This Week" at *The Economist* (London), 20–26 March 2010, p. 10.

50 "In recent, past Sinopec has been mired with allegations of unethical activity. In response, since 2007, the company requires employees to sign anti-corruption pledges, inserts anti-bribery clauses into legal contracts, and implemented an online-based corporate procurement program". For information about this and other charges, see: J. T. Areddy, "Sinopec is Linked to Daimler", *The Wall Street Journal*, 27–28 March 2010, p. A8.

51 John Pomfret, "China Invests Heavily in Brazil, Elsewhere in Pursuit of Political Heft".

52 Economic Commission for Latin America and the Caribbean, *Preliminary Overview of the Latin American Economies* (Santiago, Chile: ECLAC, 2008), p. 42.

53 For instance, the 2014 World Cup and the 2016 Olympic Games will be held in Brazil. These "political" victories can be seen as announcing the country's arrival as an accountable and respectable international player.

54 For an extended essay, see: "The EU-Latin America Summit: plus ça change", *The Economist*, 13 May 2010.

55 Some other facts about Brazil's prestige include the world's largest cattle industry, the world's greenest economy, a sophisticated bio-fuel industry, a modern financial hub (in São Paolo), deriving 80% of energy needs from hydropower, hosting 14% of fresh water in the world, a privileged abundance on natural resources, nutrient-rich arable farmlands and a sustained average GDP growth of 7%, which is three times that of the USA.

56 Peter Hakim, "The Reluctant Partner", *Foreign Affairs*, Vol. 83, No. 1 (January/February 2004), pp. 114–23.

57 He Li, "China's Growing Interest in Latin America and its Implications", *Journal of Strategic Studies*, Vol. 30, No. 4–5 (2007), p. 834; "Rethinking the Third World: Seeing the World Differently", *The Economist*, 10 June 2010.

58 For comprehensive reviews, see: Sigfrido Burgos Cáceres and Sophal Ear, "China's Strategic Interests in Cambodia: Influence and Resources", *Asian Survey*; Sigfrido Burgos and Sophal Ear, "The Geopolitics of China's Global Resources Quest", *Geopolitics*; Sigfrido Burgos Cáceres and Sophal Ear, "China's Oil Hunger in Angola: History and Perspective", *Journal of Contemporary China*.

59 Main publications by UNCTAD, available at www.unctad.org/Templates/Page.asp?intItemID=1717&lang=1 (accessed 23 December 2010); "China and America: Repelling Borders", *The Economist*, 8 April 2010; Organisation for Economic Co-operation and Development and Inter-American Development Bank, "Competition Law and Policy in Brazil: A Peer Review" (Paris, France: OECD/IDB 2010), pp. 61–68.

60 In some cases, China has extended credit lines and billions of dollars in loans at less than 1% interest rate. Some loans originate at the China Development Bank Corp., the state-run bank for public works projects.

61 Over the years, the OECD Competition Committee has done significant and relevant work in the areas of abuse, dominance and monopolization, cartels and bid rigging, mergers, prosecutions and laws, and regulation of industry sectors. See the OECD website: www.oecd.org (accessed 23 December 2010).

62 G. Luft, *Fuelling the Dragon: China's Race into the Oil Market* (Washington, DC: IAGS), www.iags.org/china.

63 Hongyi Harry Lai, "China Builds Bridges to Fuel its Engine Room", *Financial Times*, 4 July 2010; Larry Luxner, "In Caribbean and Latin America, Chinese Dragon Wakes Up", *Seis Continentes* (Summer 2005).

64 Anthony Hall, "From fome zero to bolsa família: Social Policies and Poverty Alleviation Under Lula", *Journal of Latin American Studies*, Vol. 38, No. 4 (October 2006), pp. 689–709.

65 Lula da Silva, formerly a metal worker, is said to have turned the country around. His policies included an infusion of cash ($115 per household) to the lowest social strata, which resulted in lifting 21 million people out of poverty and into the middle class. The aggregate rise in disposable incomes precipitated an expansion of sales in refrigerators, air conditioners, automobiles, motorcycles, electronics and essential household items.

66 David Samuels, "From Socialism to Social Democracy: Party Organization and the Transformation of the Workers' Party in Brazil", *Comparative Political Studies*, Vol. 37, No. 9 (November 2004), pp. 999–1024.

67 For a review on energy policy, see: Michal Meidan, Philip Andrews-Speed and Ma Xin, "Shaping China's Energy Policy: Actors and Process", *Journal of Contemporary China*, Vol. 18, No. 61 (September 2009), p. 602.

68 For an in-depth study on the paradoxes and limits of power between Brazil and the USA, see: Robert Wesson, *The United States and Brazil: Limits of Influence* (Praeger and the Hoover Institution, 1981).

69 The business-friendly economic policies of the Lula da Silva administration related to growth and development, conservative fiscal policies, credit oversight and tight banking regulations are largely credited for leaving the country unscathed by the world financial crises that crept in 2008.

70 It is worth noting that Brazil possesses no nuclear arsenal or nuclear ambitions; since 1870 it has not been involved in any war and sent only a very small number of troops to conflicts waged by other powerful nations.

71 Juan Forero, "China's Oil Diplomacy in Latin America", *The New York Times*, 1 March 2005, available at www.nytimes.com/2005/03/01/business/worldbusiness/01oil.html?pagewanted=print& position = (accessed 15 July 2012).

72 Daniel P. Erikson and Janice Chen, "China, Taiwan, and the Battle for Latin America", *The Fletcher Forum of World Affairs*, Vol. 31, No. 2 (Summer 2007), pp. 69–89; Cynthia Watson, "Adios Taipei, Hola Beijing: Taiwan's Relations with Latin America", *China Brief*, Vol. 4, No. 11 (May 2004), p. 8.

73 Taiwan has formal diplomatic relations with: Paraguay, Guatemala, El Salvador, Nicaragua, Honduras, Costa Rica, Panama, Belize, the Dominican Republic, Haiti, St Vincent and the Grenadines, St Kitts and Nevis, Gambia, Burkina Faso, Malawi, Swaziland, São Tomé and Príncipe, Tuvalu, the Solomon Islands, Nauru, Kiribati, Palau, the Marshall Islands, St Lucia and the Vatican. Refer to, for example, www.bilaterals.org (accessed 12 April 2007); see also Jorge I. Domínguez, *China's Relations with Latin America: Shared Gains, Asymmetric Hopes*, (Inter-American Dialogue, China, June 2006), p. 15, available at dspace.cigilibrary.org/jspui/bitstream/123456789/21451/1/Chinas%20Relations%20with%20Latin%20America%20%20Shared%20Gains%20Assymetric%20Hopes.pdf?1 (accessed 25 July 2012).

74 Chinese reunification refers to the bringing together of all of the territories controlled by the People's Republic of China (PRC) and the Republic of China (ROC) under a single political entity.

75 In early 2010, China hinted that it might support Brazil's dream of becoming a permanent UN Security Council member, as noted in an article by John Pomfret, "China Invests Heavily in Brazil, Elsewhere in Pursuit of Political Heft".

76 Victor Bulmer-Thomas, "Latin America: Forging Partnerships in a Transformed Region", in Robin Niblett (ed.), *America and a Changed World: A Question of Leadership* (Wiley-Blackwell and Chatham House, 2010), p. 25.

77 "Brazil, Argentina, and Mexico have led a buoyant recovery from the global recession, according to Reuters", available at www.nodeju.com/3654/america-longer-global-force.html (accessed 15 July 2012).

78 BBC News, "Brazil Seeks More Access to China to Balance Trade", *BBC News – Business*, 14 February 2012, available at www.bbc.co.uk/news/business-17022432 (accessed 14 July 2012).

79 Mitch Moxley, "Beijing Strengthens Links with Latin America", *Asia Times*, China Business, 20 August 2010, available at www.atimes.com/atimes/China_Business/LH20Cb02.html (accessed 15 July 2012).

80 Amy Myers Jaffe and Steven W. Lewis, "Beijing's Oil Diplomacy", *Survival*, Vol. 44, No. 1 (Spring 2002); Kim Woodard, *International Energy Relations of China* (Stanford, CA: Stanford University Press, 1981).

81 While drafting this chapter, a change in leadership took place. Lula da Silva concluded his presidency in 2010 with a 77% approval rating (one of the most popular and loved leaders in the world). In replacement came the first female president in the history of Brazil, Dilma Rousseff, a former chief of staff handpicked by Lula da Silva.

82 Michael Reid, *Forgotten Continent: The Battle for Latin America's Soul* (Yale University Press, 2007); Jorge G. Castañeda, "Morning in Latin America", *Foreign Affairs*, Vol. 87, No. 5 (September/October 2008).

83 For more details, see: Peter Brookes, "China's Influence in the Western Hemisphere", Heritage Lectures No. 873 (Washington, DC: The Heritage Foundation, April 2005), www.heritage.org (accessed 3 September 2010); Bulmer-Thomas, "Latin America: Forging Partnerships in a Transformed Region", pp. 15–29.

84 Their relations go beyond economics: there is talk that Brazil could train Chinese pilots on its aircraft carrier. This signals mutual collaborative desires in military and geostrategic spheres that pose a threat to local hegemony.

85 Mamdouh G. Salameh, *China's Global Oil Diplomacy: Benign or Hostile?* (Cleveland, OH: International Association for Energy Economics, first quarter 2010), pp. 21–25, www.iaee.org (accessed 23 December 2010); Pomfret, "China Invests Heavily in Brazil, Elsewhere in Pursuit of Political Heft".

86 John Mitchell, *The New Oil Economy: Impacts on Business, Geopolitics and Society* (London: Earthscan, 2001).

5 Cambodia

China's strategic interests in Cambodia are part of a wider strategic agenda in Southeast Asia that is a natural outgrowth of Beijing's 1955 "Principles of Peaceful Coexistence". This document appealed for "respect of the sovereignty and territorial integrity of all states, non-aggression against other states, non-interference in the internal affairs of other states, equality and mutual benefits, and peaceful coexistence".[1] This comes as no surprise. Southeast Asia is home to roughly 525 million inhabitants and is an economically relevant trade partner to the USA, the European Union (EU) and Japan. The region accrues a gross national product (GNP) of more than US $700 billion, hosts a number of nascent democracies, and sits atop some of the world's most transited sea routes, giving it a pivotal geostrategic position from which to oversee half of global trade.[2]

Cambodia has been a compliant ally of Beijing ever since it was clear that mutual benefits existed. For China, any nation-state that actively furthers the Chinese regional strategic agenda is welcomed as a friend, regardless of who is at the helm. Evidently, China's potential utility in shaping its ally's economy argues that Beijing should continue to play a dominant role in Cambodia. Through a literature review that covers an array of scholarly and media sources, this chapter attempts to elucidate some of the Sino-Cambodian linkages and the mutual benefits that arise from them. The chapter starts with a brief account of the historical ties between the two countries. This is followed by an analysis of the mechanisms, motivations and subsidiary actions designed to solidify strategic interests, and the potential consequences of such actions.

Historical background

History tells us that "Cambodia gained independence from France in 1953".[3] Formal Sino-Cambodian relations began in 1958 immediately after the French system of administering Chinese communities was at last terminated. This colonial system levied taxes and fees on all Chinese and non-Chinese communities residing in Cambodia in order to develop economic infrastructure, just as was done in British Burma and Malaya. Between the late 1960s and early 1970s, ethnic Chinese were Cambodia's largest minority, "but by 1984, as a result of Khmer Rouge genocide, emigration", social unrest, and Vietnamese persecution "only 61,400 remained".[4]

In 1970, Cambodia's monarchy was abolished. From 1972 to 1978, Maoist China financially supported Pol Pot's Communist Party of Kampuchea (Cambodia). However, after the communist regime's demise, new and friendly ties developed through Norodom Sihanouk, the "King Father" of Cambodia.

In late 1978 Vietnamese forces invaded Cambodia. Sino–Soviet tensions arose in connection with three elements: Moscow's support for the Vietnamese military intervention, the ensuing but brief 1979 war between China and Vietnam, and the Soviet invasion of Afghanistan. Regional tensions eased somewhat afterward, as a result of changes in the wider international environment.[5] During the 1980s, a Cambodian resistance movement fought for Cambodian independence. Vietnam withdrew from Cambodian territory as China remained vigilant over geopolitical sensitivities. Hun Sen became prime minister, and the Vietnamese-backed government gained momentum against the rebels.

In June 1989, the bloody suppression of the "Tiananmen Square demonstrations in Beijing led to Western sanctions against China".[6] During the 1990s, multi-party democracy was restored in Cambodia, where a Chinese cultural revival occurred. Cambodia's relations with Taiwan were officially severed in 1996, and a coup d'état in 1997 by Second Prime Minister Hun Sen ousted First Prime Minister Norodom Ranariddh; Chinese-Cambodian lawmakers and senior officials began to occupy key power positions.[7]

From 2000 to date, the Sino-Cambodian relationship has reflected a newly evolving geopolitical landscape characterized by flexible and opportunistic arrangements that are often volatile. Today, Cambodians of Chinese ethnicity represent roughly 2.5% of Cambodia's population.[8]

Strategic interests in Cambodia

To China, exerting greater influence in the region and securing abundant natural resources (discussed below) are two of the most salient motivations to nurture a reciprocal relationship with Cambodia. Now, as in the past, these linkages are supported by shifting political and economic contexts. What is different today is China's undisputed ascension as a powerful regional and global player, aided by trade liberalization, dynamic regional development, and a favourable international economic climate. The breadth and depth of Chinese foreign policy in recent years have been motivated by concrete and fairly narrow goals such as maximizing access to overseas markets, energy sources and the necessary raw materials to sustain growth. Having a special relationship with China allows Cambodia to develop its economy, build primary and secondary infrastructure (permitting it to steer clear of conditional loans and grants from Western donors), and be perceived as a long-term, loyal ally in Southeast Asia. Although China's approach is hardly country specific, it is now evident that some countries in the region are more equal than others. This is especially true when historical ties, behavioural legacies and tactical geopolitics play out in ways that largely help fulfil China's desire to establish itself through a charm offensive as an emerging superpower.[9] To better elucidate these tactical moves, selected issues are discussed below: security and oversight, energy self-sufficiency, infrastructure projects, financial hegemony, strategic approaches and securing resources. A brief analysis follows on the consequences of intervention, focusing on labour abuse, livelihood and ecologic disruptions. A conclusion offers summary findings.

Security and oversight

Few would disagree that the traditional underpinnings of foreign relations in Southeast Asia are undergoing a dramatic overhaul owing to China's increasing presence, a relative decline of American hegemony, amelioration of antagonistic bilateral relationships, the expanding normative influence of regional multilateral institutions, and a growing sense of potentially beneficial economic interdependence among stakeholders. China's new security and military postures, and concomitant strategies, are supported by its distinctive diplomatic approach and increasing involvement in regional affairs. These activities reflect China's newfound confidence and determination in realms that for the most part had been the preserve of American and Soviet spheres of influence.[10] These approaches—some more visible than others—are flourishing in largely uncertain circumstances where classic paradigms of the balance of power interact dynamically with new economic and security challenges. These include trade globalization, energy security, financial stability, international terrorism, nuclear proliferation and the spread of emerging and re-emerging infectious diseases.[11]

As a complement to traditional security issues, a new mode of international cooperation has arisen: the field of non-traditional security. This is defined as increased attention to anti-proliferation and anti-terrorism issues. In the Chinese case, awareness of potential threats has created fertile political grounds to justify China's increasingly active presence in these shrouded and sensitive spheres. For instance, "China has played a notable role in seeking a peaceful solution to the nuclear issue on the Korean Peninsula"[12] as a member of the Six-Party Talks. This role has evolved despite Beijing's rejection by the Association of Southeast Asian Nations (ASEAN) in its effort to join the Southeast Asia Nuclear Weapon Free Zone accord in 2004. Moreover, that November, Beijing "hosted a Security Policy Conference with officials from the ASEAN Regional Forum to discuss potential future challenges to regional security".[13]

More recently, China has been active and vocal, yet always ambiguous, about Iran's and North Korea's nuclear programmes. In the past, China's position in relation to Iran and North Korea was outright supportive, but heavy Western lobbying has, somewhat, distorted China's embrace. Now, as in the past, China disputes its sea borders with several countries, most heatedly with Japan, the Philippines and Vietnam. Also, throughout their histories, Vietnam and China have engaged in conflicts on their borders as well as in their waters.[14]

In recent times, these conflicts have been narrowed to fishing and navigation rights in the South China Sea. However, discord over territories and water does not prevent China and its neighbours from developing collaborative relationships aimed at building regional cohesion, in order to foster frictionless, mutually beneficial multilateral trade. It stands to reason that rapidly improving trade relations (and economic windfalls) will mitigate any potential fallout from territorial disputes. For instance, in the past decade, improved Sino-Russian relations based on regional strategic convergence and natural resources have provided evidence supporting this utilitarian camaraderie.

In consideration of a novel, all-embracing principle of harmonious coexistence, China's domestic and regional security relations with its neighbours are essentially shaped by shifting ideological configurations, the evolution of foreign relations with other countries, and geographical proximity. Having unrestricted air, land and sea access to the eastern side of the Gulf of Thailand powerfully motivates Beijing to engage in economic diplomacy. Between 2005 and 2007 China donated nine patrol boats and five warships to Cambodia. These are now docked at the Port of Sihanoukville, about 230 kilometres south-west of Phnom Penh, Cambodia's capital. China argues that it is helping Cambodia strengthen its marine security to fight off pirates and drug smugglers, with improvements in social stability and territorial security as spill-over effects. Although this may seem in principle a good-natured gesture, it can also be seen as an overt attempt to establish a military bulwark to counteract present and future threats that may arise in the region. Should tangible future threats arise, Beijing could build on improved military capacity in Cambodia's coastal stretch off the Gulf of Thailand to squelch it, thus, granting China valuable time to deliver reinforcements as needed. The current donated ships, however, are solely operated by Cambodia.

The concern is that such threats could jeopardize Chinese interests in Cambodia, especially a tactically relevant seaport through which import deliveries take place. Geopolitically speaking, China's interest in Cambodia can easily be linked to its security and oversight issues in the South China Sea with Vietnam, Taiwan and the Philippines, as well as safeguarding "Beijing's claims to the disputed Spratly Islands and their resources (mainly oil and natural gas)".[15] For instance, if sensitive security issues were to arise in the South China Sea regarding Taiwan, Beijing may need Cambodia as a strategic, sea-accessible location from which to launch a response. Similarly, in response to recent cases of piracy and looting, the capability for rapid Chinese deployment of warships to protect its own vessels en route to the Gulf of Aden is also a means of supporting China's strategic interests. These tactical overtures signal that China's "economic support and diplomatic maturity can be conducive to security and stability" in Cambodia. By fostering bilateral cooperation and strengthening the Cambodian economy, in exchange Beijing gets to leverage its influence to mediate regional conflicts.[16]

For example, "China has played a pivotal role in mediating, through the Six-Party Talks, the nuclear crisis on the Korean Peninsula".[17] These desired outcomes are backed by relentless advocacy in order to be perceived a non-intrusive and non-threatening, especially in response to concerns among scholars and international leaders "over the potential implications of China's rise".[18] Moreover, to promote its perception as a peaceful giant in Cambodia and the region, China's foreign affairs advisers have concocted a sophisticated combination of sustained development assistance, trade incentives and confidence-building measures. Also, Beijing has been opportunistic, stepping in when American and Japanese authorities have hesitated (e.g. on aid, trade agreements and infrastructure development). By all accounts, this posture suggests an unwavering willingness to build lasting relations with neighbouring countries, where others have left off.[19] Yet, it is worth noting that practical and sudden cooperation between two countries in a populous region could be hamstrung by historical suspicions, cultural prejudices, geopolitical rivalries and evolving priorities.[20]

All things considered, it is plausible that China's security and oversight interests in Cambodia may be: 1) strongly supported by an overall collective notion of developing a region in which trade-propelled development proceeds smoothly; or 2) unlikely as it sounds, received reluctantly with silent resistance and impinged by an already solid tradition of Southeast Asian nations being "staunch supporters of the norms of sovereignty, territorial integrity, and non-intervention".[21] Whether China's strategy proves to be successful remains to be seen.

Energy self-sufficiency

The world's population is projected to grow to more than 9 billion by 2050. As of April 2010, it was estimated by the US Census Bureau that there were 6.8 billion people in the world, of which 1.3 billion reside in China (roughly 20%).[22] China's immense population is growing at a rate of 0.629% per year (or roughly 8.5 million). As a result of China's rapid population growth and economic development, an increasingly affluent growing middle class is demanding more and more products and services. The country's industrial sector desperately needs to secure inexpensive, reliable energy sources to keep the engines of middle-class consumption growth at full speed, once the world's economy recovers from the recession. China is likely the world's largest consumer of oil, followed closely by USA.

The jury is still out on whether oil reserves around the world will dwindle in the near future.[23] As with oil, China is heavily dependent on cheap coal to generate much of its electricity, consuming almost 3 billion metric tonnes of coal per year. It also derives 1% of its power needs from 11 fully functional nuclear reactors. Cognizant of these facts, China has invested heavily in developing alternative energy sources such as wind, solar, thermal and hydroelectric power.[24] Seen through an economic lens, hydroelectric power plants are the most efficient means of producing electrical energy.[25]

With this cost-effective notion in mind, China is finishing its fourth hydroelectric dam on the upper reaches of the Mekong River, sparking concerns among its neighbours about future water supplies. China has also promised to build hydroelectric dams along the Mekong inside Cambodian territory. The Cambodian authorities aim to triple Cambodia's energy output from a projected 808 megawatts in 2009 to nearly 4,000 megawatts by 2020, with the goal of connecting nearly 70% of the population to sources of electricity.[26]

Although Cambodians may be wary of these developments, there are strong incentives to proceed. The proposed Chinese dams along the Mekong River portion of Cambodia would likely create desperately needed jobs in the northeastern provinces and eventually enable potential surplus energy sales to Thailand and Vietnam. Furthermore, to China's advantage is the fact that once hydroelectric dams are in place in the upper and lower Mekong Basin, controlling water flow and energy production could easily become a political tool to advance China's long-term energy and resource plans. If nearby oil is what China is looking for, then securing long-term oil reserves should be uncomplicated because the "Cambodian National Petroleum Authority, which administers oil contracts, is under the direct control of a prime minister and deputy prime minister"[27] all too happy to sing China's praises.[28]

China has allegedly acquired the rights to offshore Cambodian oil fields by purchasing a firm with pre-existing claims. Recent geological work suggests that significant amounts of oil and gas may reside underground throughout Cambodian territory. These reserves are often hailed as providing a way out of poverty.[29] It should be noted, however, that proven reserves are scarce, and tapping them is expensive, given prevailing energy prices, high exploration and production costs, and low levels of technology. In practice, Chinese officials weigh higher demand against lower reserves, perceiving their existence as an exploitable opportunity. Whether China's interest is oil, hydroelectric power, or both will soon be elucidated as its strategy unfolds. China, either alone, or in conjunction with its Southeast Asian partners, aspires through decentralized energy generation structures to become energy independent.

This lofty strategy may backfire if mainland Southeast Asian nations conclude that Chinese dams, oil exploration or any other gargantuan infrastructure projects are hurting their future development prospects. Some dispersed action groups, civil society organizations and environmental activists are already protesting China's dams and oil thirst.[30] What is now undisputed is that energy self-sufficiency is a central goal for China, one that can be attained through a multi-dimensional action plan[31] that places Cambodia as an important partner.

Infrastructure projects

As implied above, the most important infrastructure projects embarked on by China are related to development of the Mekong River through building several hydroelectric dams.[32] China's strategic advantage in water resource management arises from its upstream contribution to Mekong waterways. For instance, it is estimated that during the rainy season, 20% of Mekong waters originate in China, while during the dry season its contribution jumps to 70%. In addition to building hydroelectric dams along the Mekong, many other projects are envisaged to cope with the logistics of China's future economic development. For example, a big initiative will modernize Sihanoukville's port and its associated facilities for import and exports of goods—and also to host a naval presence.[33] Naturally, goods will require transport to and from Cambodian cities; therefore, establishing a series of roads, highways and bridges is a subsidiary development project. "In 2006, China offered Cambodia $600 million in no-strings-attached loans to help pay for two major bridges near Phnom Penh that will link into a network of primary and secondary roads, and also to pay for a modern fibre-optic network to connect Cambodia's telecommunications with Vietnam and Thailand".[34] Even construction of modern railroads connecting major cities to the port has been proposed.[35]

Chinese projects in Cambodia are not limited to hydroelectricity, ports, roads, highways and bridges. As with East Timor's presidential palace, China's infrastructure projects also include prestigious buildings such as Cambodia's Senate headquarters (completed in 1999) and the recently finished Council of Ministers building for which China sent work crews to labour 24 hours a day. Prime Minister Hun Sen refused to move into his office, allegedly because of feng shui (geomantic) concerns. He ordered the construction of an annex next door—perhaps also to avoid the possibility of pre-installed listening devices. Large mining and forestry projects, oil exploration, biofuel generation, telecommunications and manufacturing have also enjoyed Chinese patronage—to Cambodians, surely, new industries such as "oil, gas, minerals, forestry, and hydro-energy offer lucrative financial rewards and new employment"[36] opportunities.

Cambodia is not an isolated recipient of this benevolence: Beijing has loaned other Southeast Asian nations large sums of money for railways, hydropower projects and shipbuilding yards. In the past five years it has signed declarations proclaiming strategic partnerships with Indonesia and the Philippines.[37] Although all "proposed infrastructure projects in Cambodia have aroused controversy" and sparked harsh criticism, China's plan "to build dams across the lower Mekong Basin has ignited the most fervent debate".[38]

Financial hegemony

The majority of infrastructure projects envisaged in Cambodia will be jointly financed by Chinese financial institutions and the Chinese Government itself. The robust financial support for these large-scale enterprises represents a surreptitious windfall for Cambodian officials acting as gatekeepers. However, ordinary Cambodians are also expected to see an influx of fresh funds flooding their fragile yet bustling economy. In the past 10 years, the Cambodian

Government has struggled to diversify its fledgling economic engine "away from reliance on tourism, low-cost garment manufacturing, and foreign aid. Thus, China's growing investment in and generosity toward Cambodia come at a critical time".[39] Within the region, China is exerting preponderant influence through financial absolutism, wielding yet another tool to coax Southeast Asians and Cambodians to fall into line.[40]

Although Beijing's giant state corporations have invested billions of dollars in dams, oilfields, highways, textile operations and mines, it is the low-profile, family-owned Chinese firms that have come to dominate the Cambodian investment landscape. For instance, in 2003 Chinese investment in Cambodia totalled $45 million. This was followed by $90 million in 2004, when the Cambodian Investment Board—heavily influenced by Chinese lobbying—approved more investment from China than from any other country. As a result, in 2005 total investment increased to roughly $450 million and to approximately $600 million in 2006. For fiscal year 2005–06, overall bilateral trade amounted to $732 million. In 2006–07, it reached $842 million—up 15% from the previous year. Higher bilateral trade and Chinese investment amounts are expected in future. For example, in December 2009 Chinese Vice President Xi Jinping inked $1.2 billion in aid and investments to Cambodia, more than was subsequently committed to the country by all official donors to Cambodia that year and each year to date. More recently, in the aftermath of the global recession, the International Monetary Fund (IMF) predicted foreign direct investment in Cambodia would drop to $490 million in 2009 from $850 million in 2008, and indeed there was a slowdown investment-wise and economically. However, despite the slowdown, some economic sectors shine brighter than others. Agriculture, for example, has fared better than the garment industry and tourism, which both retrenched following the global financial crisis.[41]

Cambodia's accession to the World Trade Organization (WTO) in 2004, carried out to preserve Cambodia's access to the global trading commons, especially for its garment sector (14% of gross domestic product—GDP) with the end of the Multi-Fibre Arrangements, has not prevented retrenchment in that sector. As part of a wider initiative on food security and development, investment in rural roads and irrigation systems is being encouraged to raise productivity and reduce operating costs; and it has helped to produce good crop harvests.[42]

To further attract international investors, the Council for the Development of Cambodia, the Government's chief investment body, is keen to approve foreign direct investment in agricultural and industrial sectors. This agency, in collaboration with the Ministry of Commerce, actively promotes the establishment of special economic zones (SEZs) such as the Phnom Penh Special Economic Zone (PPSEZ). This zone was established in 2006, as enclosed, tax-exempt, complex-partnered and co-financed by Singaporean, Japanese, Malaysian and Chinese private firms. The PPSEZ is one of 19 SEZs in Cambodia. It is expected that upstream and downstream businesses, and their associated networks, will greatly benefit from SEZs even if they are not located within the confines of an SEZ. This will not only give China increased political leverage over Cambodian officials but will grant the Chinese a stronger foothold in negotiations with the ruling Cambodian People's Party, given that follow-up investments in phased projects can be strategically withheld or halted if regional or domestic policies do not converge as desired by Beijing.[43]

To sweeten many of these deals, China has already agreed to waive import tariffs on almost 400 Cambodian products. At the same time, and equally important, China envisages a bilateral free trade agreement with Cambodia. Despite an avalanche of bilateral benefits accruing on both sides, there are fears that latent resentment may surface against a sizable population of Chinese-Cambodian economic elites, owing to a growing Chinese economic presence. In some sectors such as furniture, garments and perishable produce, China's expansion is not welcome because it supplants those traditionally produced locally.

China not only encourages Chinese investments in Cambodia but also finances established Chinese-Cambodian firms that make cost-effective usage of inexpensive and plentiful local labour. As appealing as China's financial domination may appear to officials in Beijing and Phnom Penh, there will always be unexpected situations. With this in mind, it stands to reason that Chinese political strategists are busy planning for contingencies.

Strategic approaches

An alternative interpretation of China's emerging might and friendlier diplomacy is what some scholars call soft power—that is, the "application of China's expanding economy to trade, aid, and investment"[44] to achieve political ends: a case in point is China's largesse in the Southeast Asian region.[45] Chinese development assistance and aid flows freely to Laos, Myanmar and Cambodia. Especially in Cambodia, China acts as a trading partner and donor, with an eye to buffering Vietnam's and Taiwan's geopolitical leverage in the region, which derives from their international political buy-in (i.e. approbation or influence), their impressive economic growth and their access to sea borders.[46]

It is argued, in scholarly and journalistic circles, that China's application of diplomatic overtures, military initiatives and economic cooperation proposals are merely instruments that regulate power relationships with "friendly" countries. China's goal is building solid allegiances to advance its agendas related to foreign policy, image shaping, economic pre-eminence and national security. All these instruments play out in Cambodia. The structure of China's soft power engagements with Cambodia takes two forms: 1) multilateral cooperation within institutional bodies; and 2) bilateral cooperative agreements. Multilateral cooperation has been particularly strong in the past four years. In 2006 combined loans to Cambodia from donors totalled $601 million, of which China contributed over a one-tenth via indirect, informal networks. In 2007 China for the first time formally pledged aid of $92 million, a

15% increase from the previous year, as part of the Development Partners (e.g. Organisation for Economic Co-operation and Development—OECD—member countries) package totalling $689 million. In 2008 international donors pledged some $952 million, with China leading the way with $257 million, followed by the EU ($214 million) and Japan ($113 million).[47]

For comparative purposes, in 2008 American aid commitments to Cambodia reached just $57 million, a little over one-fifth of China's commitment (which is invariably higher than actual disbursements). As for bilateral cooperative agreements, these are harder to estimate, given the complexity of transactions that overlap across time and sectors. None the less, some data exist: in 2006 and 2008 China alone committed $600 million and $215 million, respectively, for infrastructure and development projects.[48] In a recent interview, Prime Minister Hun Sen asserted that "from 1992 to June 2009, China's financial aid to Cambodia totalled $923 million".[49] It is safe to say that Beijing's sugar-coated relationship with Cambodian officials has thickened.[50]

By disbursing aid and investment funds liberally and without conditions, the donor country makes the recipient country heavily dependent on it. In the process, China fends off other lenders and donors who impose complicated restrictions on their largesse. The Chinese approach, it can be argued, provides political leverage that senior officials can exploit in pressing forward their economic and political ends. To illustrate this: in February 2009 Chinese Foreign Minister Yang Jiechi pledged $55 million in aid and $1 billion in direct investment to develop aggressively Cambodia's power industry, while vowing to increase bilateral trade beyond $750 million. These lofty promises undermine an already weak governance capacity in Cambodia and could prompt unrealistic expectations, strengthened patronage networks and greater rent-seeking behaviour. Chinese aid for infrastructure projects targets trade integration in the region, which is, by all accounts, a useful outcome for poor countries with weak governance.

In December 2009 20 Uighurs somehow made their way to Phnom Penh and sought refugee status from the UN High Commissioner for Refugees, but "were deported back to China the day before Chinese Vice President Xi Jinping visited Cambodia"[51] and signed $1.2 billion in aid and investment deals. A few months later, the USA cancelled plans to donate 200 surplus military trucks to Cambodia. China responded by donating 257 military trucks and 50,000 uniforms. Then, when the USA announced it would not forgive $300 million owed from the pro-American Khmer Republic, China forgave Cambodia's Khmer Rouge era debt of several million dollars. These are but a few ways in which China is working with its Southeast Asian neighbours to become more palatable, and to borrow influence from the USA.

Furthermore, a palatable approach for the Cambodian private sector is the opening of Chinese markets for their products, services and investments. However, comparing the two economies, it is clear that Cambodia would benefit more from China than China from Cambodia. In terms of markets, for instance, Cambodia nets a win-win situation because it gains access to larger markets for its products, as well as faster development of its manufacturing industries thanks to Chinese know-how. Nevertheless, opportunities will surely abound for both countries. For instance, in Cambodia alone there are roughly 3,000 Chinese-registered companies operating, most of them related to garments and textiles. In 2007–08 there was interest for the Cambodian export of rice to China, but the latter scuttled one deal when it found phytosanitary issues related to bacterial contamination. Most important to China is open and easy access to Sihanoukville's port, which will in effect guarantee deliveries of imported oil and mobilization of resources as needed. It is important to highlight, in the context of Sino-Cambodian affairs, that the "China-ASEAN Free Trade Area began on January 1, 2010", with the goal of a zero tariff rate for 90% of goods traded among China, Cambodia, Laos, Myanmar and Vietnam by 2015.[52]

Another strategic approach being used can be best defined as cultural force-feeding. In its most basic form, this means the use of cultural background to influence states and peoples in other locations. In Cambodia, this takes form as cultural showcasing and the promotion of the Chinese language. In 1990, through political coaxing, senior Cambodian officials permitted the establishment of multilingual institutions; Chinese dialects grew in prominence. China contributes to this cultural and educational rubric by funding school construction, textbook production and training of teachers, as well as offering scholarships to Chinese-Cambodians for secondary and university education in China. It should be noted that not only Cambodians of Chinese ancestry study Chinese, but pure ethnic Khmer do too, realizing the advantages of learning the language and customs of one of Cambodia's most powerful partners. Another influential front has been the establishment of Chinese-language newspapers in Cambodia, coupled with Chinese television entertainment and news broadcasts on cable TV channels.

Similarly, Beijing has encouraged and supported Chinese-Cambodian social organizations (e.g. the Cambodian-Chinese Association, based in Phnom Penh, and clan-based organizations), many of which promote Chinese roles, practices and demeanours to young Cambodian enthusiasts. As a result of the near decimation of ancient Cambodian cultural heritage by the Khmer Rouge regime from 1975 to 1979, China spotted a golden opportunity to support aggressively a Chinese cultural revival. This cultural and educational expansion results in wider acceptance among Cambodians of Chinese ways, ideologies, policies, products and services.[53]

The regionally diverse portfolio of Beijing's engagement strategies and diplomatic approaches suggests that not all Southeast Asian countries share a uniform perception about China's intentions. Each country has its own unique relationship with China, reaps its own benefits, and deals with its own difficulties. Although a great deal of money has been doled out throughout the region, China's special generosity toward Cambodia has caught the attention of Western governments[54] and caused alarm over the potential tilt in the region in favour of a non-traditional hegemony.

Securing resources

High food prices, rapidly emerging animal and plant pests and diseases, and policy-induced supply shocks all ignite fears among Chinese officials that dependence on world food markets to satisfy domestic demand would leave China vulnerable. As a result, leaders realize the need for access to critical commodities, not only to feed an expanding population, but also to meet the demands of insatiable output and export-oriented industries.

More broadly, growing concerns about future scarcity have led foreign governments to undertake large-scale land acquisitions abroad. A multi-institutional study indicates that in agriculture-oriented countries, land-based investments have risen over the past half decade, spurred by foreign investment that seeks to exploit opportunities for its own advantage.[55]

China is thirsty for natural resources such as timber, gas, oil, water, rubber, fertile cropland and minerals (i.e. gold, silver and iron-ore). More importantly, it is keenly interested in having cheap, unprotected, unrepresented and readily available labour forces abroad.

This interest reflects the fact that the great majority of Chinese exports are still labour intensive. The main comparative advantage of Chinese manufacturing power lies in its abundant labour supply, but its wages have inevitably risen in concert with economic growth. To witness this manufacturing phenomenon, investors and purchasing managers are encouraged to visit the city of Yiwu, also commonly referred to as the International Trade City, southwest of Shanghai, which hosts what is alleged to be the biggest market in the world, the Yiwu Market. Here, stalls display thousands of items to potential buyers.[56] Nowadays, China faces a growing dependence on imports of important minerals and is already a crucial world market for copper, iron-ore, platinum, aluminium and precious woods.

In 2004 global production of crude steel reached the billion-ton mark almost entirely as a result of intense Chinese demand. It is no secret that China already consumes more steel, grain and coal than the USA. As if this were not enough, "China is also one of the world's largest importers of tropical timber from the region".[57] Because smaller households consume more resources per person, China's rapid increase in household numbers and reduction in household size results in rapacious resource usage.[58]

Although China has shown signs of assuming greater regional and international obligations, and is steadily becoming a responsible stakeholder, its approach to domestic and global affairs remains largely utilitarian.

Intervention consequences

For nations, there are often direct and indirect consequences related to specific interventions (mainly through infrastructure projects) which prove to be detrimental in multiple dimensions for the short, medium and long term. In the past 50 years, Cambodians have experienced—and most fear they will continue to experience—a series of negative–impact actions: unlawful land-grabbing, choking of freedom of expression (not to mention human rights), illegal logging, unregulated mining, labour abuses, illicit resource exploitation and environmental damage to rivers, lakes and water-dependent ecosystems.[59]

In many cases, the Government itself has been behind these abuses. Because of the frailty of Cambodia's democracy, the profound lack of sensible political institutionalization, weak social cohesion and the utter disregard for people's rights, concern has grown over the potentially pernicious outcome of China's actions which are embraced, accepted and encouraged by Cambodian officials.

Lately, close ties between top Chinese officials and Cambodian Prime Minister Hun Sen have played a pivotal role in advancing China's strategic interests in both Cambodia and the region.[60] This is no novelty: corruption, cronyism and nepotism are endemic in Cambodia.[61]

Labour abuses

Cambodians have a long history of suffering human rights and labour abuses.[62] In response to allegations and in collaboration with the Royal Government of Cambodia and the International Labour Organization, the USA and the EU have advanced innovative, incentive-based policy experiments (e.g. expanded export quotas as incentive), which have rendered positive results in garment factories,[63] but there is much room for improvement in other industries.

Some of the most common features of abuse and exploitation, anecdotally reported, are harsh and demeaning working conditions, unpaid overtime, harassment, unauthorized wage deductions, underage workers, unpaid sick leave, verbal abuse, delayed compensation and suppression of workers' complaint mechanisms. In many cases, because of a profound lack of gainful alternative jobs, many Cambodians agree to abusive and exploitative work conditions, lest they join the ranks of the unemployed (known colloquially as *dah twat ktiol*—literally, walk-kick air). Recently, the World Bank forecast that 200,000 Cambodians would fall below the poverty line in 2009, partly as a result of the global economic slowdown.[64]

It is feared that young, unemployed women will turn to prostitution or begging to make up for household budget shortfalls. Households facing multi-dimensional insecurities could try to cope with income losses by migrating to nearby cities, selling valuable assets such as livestock, borrowing money from usurers, or encouraging their young children to work.

Livelihood and ecological disruptions

In the past quarter of a century, international organizations have come to recognize that sale or long-term lease of farmland to foreign governments or international investors can result in land expropriation.[65] Alternatively, it may ultimately lead to unsustainable resource use, thereby undermining the livelihoods of local communities through job loss or income reduction, in addition to other pernicious outcomes. However, the other side of the argument is that land sales or transfers can also bring opportunities to local communities, including employment creation and demand-driven multiplier effects through backward and forward network linkages.[66]

Ecologists and environmentalists, aided by classical and contemporary tools to measure anthropogenic effects[67] on ecosystems, have highlighted the loss of productive land to non-productive uses, sometimes leading to slash-and-burn deforestation. Often, in-depth investigations with farmers in rural areas reveal that continuous farming of these in-transition plots reduces crop yields. This stems from factors such as soil erosion, inadequate pest control practices and low-quality inputs (seeds, pesticides and fertilizers), among other things. Similarly, non-governmental organizations (NGOs) have denounced unrestrained mining exploration in northern Cambodia that has damaged six of the country's 23 protected wildlife areas. Loss of livelihood based on bamboo harvesting and resin tapping from trees—because of extensive logging—has also been reported recently.[68]

Because no country in the region is more reliant on the Mekong River than Cambodia, English-language local newspapers such as the *Phnom Penh Post* and the *Cambodia Daily*, as well as some Khmer-language newspapers, have been covering river-related issues. They have reported stories of villagers in river-side communities forced to move because the flooding Mekong destroyed their lands. Rural and indigenous communities that depend on animal hunting, fruit gathering, fishing, or water transport along the Mekong River and the Tonle Sap (Cambodia's great lake) have voiced fears that Chinese-backed infrastructure developments will damage the environment and threaten their livelihoods.[69]

In the environmental arena, China is perceived as the source of numerous regional ecological disruptions. China's dam development has already hurt Cambodia. By mid-2004 the Mekong water level had reached record lows, resulting in a nearly 50% drop in the fish catch from the previous year and also menacing rice production. It is equally alarming how many households could experience drops in tourism-related income in areas around the Tonle Sap and Siem Reap (home to the famed Angkor temples), which potentially could push people into food insecurity, mass urban migration, prostitution, increased violence and protests.[70]

Social anthropologists have warned of the unexpected increase in costs associated with post-displacement minority group relocations. They have also pointed to the long-term developmental and health effects of diminished nutritional profiles in infants, children and adults resulting from reduced consumption of animal proteins. A 2009 report by the World Wildlife Fund (WWF) reveals that pollution in the "190-kilometer stretch of the Mekong River between Cambodia and Laos has pushed the local population of Irrawaddy dolphins to the brink of extinction while nearly decimating other endangered species and harming aquatic populations".[71]

In 2010 plans to develop the Mekong were viewed by regional countries and the international community with trepidation. "The Mekong River

Commission (MRC) was formed on April 5, 1995, by an agreement among the governments of Cambodia, Laos, Thailand, and Vietnam".[72] "The Agreement on Cooperation for the Sustainable Development of the Mekong River Basin" sought joint management of the four countries' shared water resources and development of the river's economic potential.[73]

Interestingly, China and Myanmar are only dialogue partners, not commission members. In subsequent years, the MRC hosted and encouraged multi-stakeholder consultations and roundtables better to address the most pressing issues surrounding development of the Mekong River. As a result, many detailed plans and strategies were showcased. However, despite drafting comprehensive environmental-impact assessments and mitigation plans, backed by significant funding assistance and institutional support, little improvement has been seen. Among other factors, this derives from China's determination to move ahead regardless of impediments, coupled with weak enforcement of environmental regulations, lack of popular condemnation, and state permissiveness. Indeed, the decimation of Cambodia's natural resource base is alarming, considering that 81% of the population of 14 million is rural, with agriculture accounting for two-fifths of GDP and constituting the main domestic activity of rural households.[74] Recent cases of highly pathogenic avian influenza (bird flu) were a valuable lesson on how a disease with global public health implications was allowed further to damage rural livelihoods via a lethal combination of international incentives and domestic politics.[75]

If this were not enough to raise eyebrows, conservationists warn that parts of the Mekong region—mainly Cambodia, Laos and China's Yunnan Province—are particularly vulnerable to the effects of anthropogenic climate change, which in all likelihood will transform sensitive ecosystems.[76] For instance, in 2004 Chinese logging operations abroad were denounced for engaging in illegal activities that spurred severe deforestation in Myanmar and Indonesia.[77] The result was soil erosion, loss of biodiversity and lower water-generation capacity. Between 2005 and 2007 civic groups emphasized the immediate need for concerted action to stop dam development along the Mekong before further human rights and environmental abuses took place.

In 2008 an International Rivers report warned of significant social and environmental costs accruing from poorly planned Chinese hydroelectric power plants being developed in Cambodia.[78] More recently, in 2009 a joint publication by the UN Environment Programme (UNEP) and the Asian Institute of Technology (AIT) urged officials to attain a balance between resource exploitation and maintenance of ecological health.[79] China's projects are now considered major contributors to the precipitous decline in biodiversity in Cambodia. In addition, tigers, bears, seahorses and rare turtles, among other exotic wildlife, are eagerly sought by some Chinese traders. The growing purchasing power within certain social groups provides strong economic incentives for the illegal wildlife trade. With all this in mind, and based on empirical evidence, it is unlikely that China will soon assume a proactive leadership role in promoting labour rights protection, fostering good governance and catering to the environment.[80]

Conclusion

From its portfolio of approaches planned and implemented so far, it is clear that China's strategic interests go far beyond projecting greater influence in Southeast Asia and securing vast supplies of natural resources in Cambodia. Chinese political and economic analysts are attuned to contemporary issues, as demonstrated through Beijing's tactical moves in Southeast Asia, most especially in Cambodia.[81] There, a less than independent judicial system[82] provides no substantial relief to ordinary Cambodians, but allows for inequities and iniquities with little potential for liability. The reality is that China has assumed a leadership role in Southeast Asian economies and aggressively pursues free trade agreements as a direct result of stagnant or declining regional trade with Europe and USA.[83] Under these circumstances, which policy options Cambodian authorities embrace to foster social stability will depend on the presence or absence of political will.[84] Policy initiatives (such as contract transparency) that harm patronage networks have the least likelihood of success given the country's political-economic landscape.[85] Naturally, officials in Beijing and Phnom Penh see this as a win-win situation. Despite much praise for its various regional and international initiatives, China has also drawn wide condemnation for failing to acknowledge the severity of Tibetan civil unrest and downplaying the Severe Acute Respiratory Syndrome (SARS) epidemic. Militarily, it is clear that the port of Sihanoukville could serve as a strategic location from which to project Chinese maritime power into the Gulf of Thailand and the Straits of Malacca, both conduits for heavy trade. Airfields in mainland Cambodia can be used as refuelling stations for military planes conducting sea and land surveillance and reconnaissance missions. Clearly, the rationale supporting a China-Cambodia military partnership is often tactical and instrumental, rather than being based on long-term, sustainable development. This status reflects the transitory nature of the contemporary international system. Diplomatically, China has improved its standing with the international community through friendly foreign policies.[86] By adopting a similar position to back Russia, Iran and North Korea on sensitive issues, Beijing can also support Cambodia when Phnom Penh is pressured by international bodies on allegations of human rights abuses, oppression, corruption and misuse of power at high government levels. Therefore, it is not surprising that the Chinese embassy is the biggest diplomatic mission in Cambodia's capital, or that China is by far Cambodia's most generous development partner. Cambodia, to repay its newfound ally for its largesse, has agreed silently to serve almost every Chinese desire and whim, from making gargantuan land concessions to allowing the building of hydroelectric dams, roads, bridges, railways and telecoms systems, as well as repatriating Uighurs and allowing China free access to its economy. For aid,

"Cambodia can always turn to China, a generous donor that despite its superpower status, seems to treat its partners with fairness and respect, never asking sensitive questions".[87] It is up to Cambodia to make the best use of these economic boosters and its intimate relationship with its giant neighbour, in order to advance economic programmes and social projects. In the long term, these could benefit Cambodia's populace and its economy, as well as cementing in place the primary and secondary infrastructure needed to sustain commerce and development. Economically, Cambodia and China will always maintain a symbiotic relationship, owing to the large number of Chinese-Cambodians. However, this relationship will mainly be reciprocally fed by the ties and benefits that bind senior officials in both countries. China's heavy investment in Cambodia stems not only from Beijing's strategic interests in the region but also arises because private-sector initiatives seek fertile ground where there is an environment to prosper and grow. Other countries have eyed the kingdom of Cambodia as a partner, be it as their vegetable garden or their rice bowl. The USA and Europe have long been a presence in the garment industry; Japan has been donor and adviser; and Australia has had a role in agricultural and industrial development. Still, Cambodian officials argue in interviews and communiqués that what is good for China is good for Cambodia, and this belief, glued by trust and respect, seems absolute. Yet, despite all the heralded advantages behind this strategic partnership, rising concerns about labour abuses, geopolitical backlashes, livelihood disruptions, social unrest, ecological perturbations, economic dependence and biodiversity losses are already making headlines. Lastly, the strategic partnership between China and Cambodia has provoked polarized views among Western observers, neighbours and the international community because it is seen as an alliance of authoritarian states that threatens ASEAN cohesion, regional stability and democratic norms. However, seen strictly through a public relations lens, China's rise has been peaceful and tactical, achieved through enhancement of global security, promotion of peaceful trade and addressing transnational challenges. Whether the outcome is positive or negative depends in large part on external forces playing out dynamically in the international arena. These include a debilitating global recession; China's willingness to play banker to America; rising nuclear and viral threats; climate change and its effect on living ecosystems; trade liberalization bottlenecks; reduced consumer confidence; and rising global poverty, inequality and hunger. Perhaps we are seeing an active redefinition of duties and responsibilities as superpowers snatch force and influence from each other, especially after traditional leadership roles have come into question around the globe.

Acknowledgement

This chapter is adapted from a refereed article entitled "China's Strategic Interests in Cambodia: Influence and Resources", by the authors, and published in *Asian Survey*, Vol. 50, No. 3 (May/June 2010), pp. 615–39.

Notes

1 Sophie Richardson, *China, Cambodia, and the Five Principles of Peaceful Co-existence* (New York: Columbia University Press, 2009), p. 9; Paul Marks, "China's Cambodia Strategy", *Parameters*, Vol. 30, No. 3 (Autumn 2000), pp. 92–108.
2 Council on Foreign Relations, *The United States and Southeast Asia: A Policy Agenda for the New Administration* (New York: Report of an Independent Task Force, 2001), pp. 1–2.
3 Cambodia, *Background of the Conflict*, n.d., available at www.preventorprotect.org/overview/cambodia.html (accessed 15 July 2012).
4 Khmer Yoeng, "Khmer Rouge Rule", *Cambodian Culture & History of Khmer*, 16 December 2010, available at khmeryoeng.wordpress.com/tag/khmer-rouge-rule/ (accessed 15 July 2012); and Wikipedia, *Racism in Asia*, n.d., available at en.wikipedia.org/wiki/Racism_in_Asia (accessed 15 July 2012).
5 Bobo Lo, *Axis of Convenience: Moscow, Beijing, and the New Geopolitics* (Washington, DC: Brookings Institution Press, 2008), p. 27.
6 Suisheng Zhao, "Nationalism's Double Edge", *The Wilson Quarterly*, Autumn 2005, available at www.gees.org/documentos/Documen-842.pdf (accessed 15 July 2012).
7 Aside from Paul Marks, "China's Cambodia Strategy", *Parameters*, Vol. 30, No. 3 (Autumn 2000), pp. 92–108; and Sophie Richardson, *China, Cambodia, and the Five Principles of Peaceful Co-existence* (New York: Columbia University Press, 2009); see: Caroline Hughes, *The Political Economy of Cambodia's Transition, 1991–2001* (London: Routledge-Curzon, 2003); Caroline Hughes, *Dependent Communities: Aid and Politics in Cambodia and East Timor* (Ithaca, NY: Cornell Southeast Asia Program, 2009); and Russell R. Ross, *Cambodia: A Country Study* (Washington, DC: Government Printing Office for the Library of Congress, 1987).
8 Amy Tsui and James Tollefson, *Language Policy, Culture, and Identity in Asian Contexts* (London: Routledge, 2006).
9 Elizabeth Economy and Michel Oksenberg, *China Joins the World: Progress and Prospects* (New York: Council on Foreign Relations Press, 1999).
10 Aaron L. Friedberg, "Ripe for Rivalry: Prospects for Peace in a Multipolar Asia", *International Security*, Vol. 18, No. 3 (Winter 1993/94), pp. 5–33.
11 P. Dibb, *Toward a New Balance of Power in Asia* (Oxford: Oxford University Press, 1995); S. Ear and S. Burgos Cáceres, "Livelihoods and Highly Pathogenic Avian Influenza in Cambodia", *World's Poultry Science Journal*, Vol. 65, No. 4 (December 2009), pp. 633–40.
12 "Proliferation from North Korea and Iran – Will Russia and China Pitch In?" *The Economist (London)*, 10 September 2009.
13 David Shambaugh, "China Engages Asia: Reshaping the Regional Order", *International Security*, Vol. 29, No. 3 (Winter 2004/05), pp. 64–99, p. 88.
14 Ang Guan, "Vietnam-China Relations Since the End of the Cold War", *Asian Survey*, Vol. 38, No. 12 (December 1998), pp. 1122–23.
15 "Q&A: South China Sea Dispute", *BBC News – Asia Pacific*, 27 June 2012, available at www.bbc.co.uk/news/world-asia-pacific-13748349 (accessed 15 July 2012).
16 Sophie Richardson, *China, Cambodia, and the Five Principles of Peaceful Co-existence* (New York: Columbia University Press, 2009); Joshua Kurlantzick, *Charm Offensive: How China's Soft Power is Transforming the World* (New Haven, CT: Yale University Press, 2007), p. 6.
17 Dick K. Nanto and Mark E. Manyin, *China-North Korea Relations*, Congressional Research Service, 28 December 2010, available at www.fas.org/sgp/crs/row/R41043.pdf (accessed 15 July 2012).
18 Elizabeth Economy, "China's Rise in Southeast Asia: Implications for Japan and the United States", *The Asia-Pacific Journal – Pacific Focus*, 6 October 2005, available at www.japanfocus.org/-Elizabeth-Economy/1815 (accessed 15 July 2012).

19 Elizabeth Economy, "China's Rise in Southeast Asia: Implications for the United States", *Journal of Contemporary China*, Vol. 14, No. 44 (August 2005), pp. 409–25.
20 Bruce Vaughn and Wayne Morrison, *China-South East Asia Relations: Trends, Issues, and Implications for the United States* (Washington, DC: Congressional Research Service, Report for Congress, April 2006), p. 4.
21 Elizabeth Economy, "China's Rise in Southeast Asia: Implications for Japan and the United States", *The Asia-Pacific Journal-Pacific Focus*, 6 October 2005, available at www.japanfocus.org/-Elizabeth-Economy/1815 (accessed 15 July 2012).
22 US Census Bureau, Data Finders, www.census.gov (accessed 18 April 2010).
23 M. A. Adelman, "The Real Oil Problem", *Regulation*, Vol. 27, No. 1 (Spring 2004), pp. 16–21; Diana Farrel, Susan Lund, Eva Gerlemann and Peter Seeburger, *The New Power Brokers: How Oil, Asia, Hedge Funds, and Private Equity are Shaping Global Capital Markets* (San Francisco: McKinsey Global Institute, October 2007).
24 Terry Macalister, "China Puts £60bn into Energy", *The Guardian*, 5 February 2009.
25 Hydroelectric power, available at www.usbr.gov/power/edu/pamphlet.pdf (accessed 18 April 2010).
26 Khouth Sophak Chakrya and Sebastian Strangio, "Villagers Gather to Express Dam Concerns", *Phnom Penh Post*, 24 September 2009.
27 "Country For Sale, Welcome to Cambodia: How Cambodia's Elite Has Captured the Country's Extractive Companies", Global Witness, February 2009, pp. 6, 7, 36, 37, available at www.globalwitness.org/sites/default/files/import/country_for_sale_high_res_english.pdf (accessed 15 July 2012).
28 "Country for Sale", pp. 1–72.
29 "Cambodia's Oil Resources – Blessing or Curse?", *The Economist*, 26 February 2009, www.economist.com/node/13184945 (accessed 15 July 2012).
30 David C. Kang, *China Rising: Peace, Power, and Order in East Asia* (New York: Columbia University Press, 2007), pp. 126–52.
31 Robert Ebel, *China's Energy Future: The Middle Kingdom Seeks its Place in the Sun* (Washington, DC: Center for Strategic and International Studies, October 2005), p. 77.
32 "Cambodia, Kuwait, and Farmland – Petrodollars v Smallholders", *The Economist*, 23 April 2009.
33 China is also reported to be interested in establishing naval bases in Bangladesh, Sri Lanka, Myanmar (Burma), Pakistan and Thailand to protect its maritime supply routes from the Middle East and Africa. See: Craig Nelson, "China's Naval Base Proposal May Raise Suspicion", *The Nation* (Abu Dhabi), 2 January 2010.
34 Jane Perlez, "China Emerges as Major Player in Asian Aid, Asia-Pacific", *The New York Times*, 17 September 2006, available at www.nytimes.com/2006/09/17/world/asia/17iht-aid.2840899.html?pagewanted=all (accessed 15 July 2012).
35 Thomas Bradley, "China's Cambodian Hegemony", *The Diplomat* (Tokyo), 7 May 2009.
36 "China's Cambodian Hegemony", *The Diplomat*, 7 May 2009, available at thediplomat.com/2009/05/07/chinas-cambodian-hegemony/ (accessed 15 July 2012).
37 Richard Halloran, "China's Paths to Hegemony", *Washington Times*, 17 January 2008.
38 "China's Cambodian Hegemony", *The Diplomat*, 7 May 2009, available at thediplomat.com/2009/05/07/chinas-cambodian-hegemony/ (accessed 15 July 2012).
39 Ibid.
40 Brendan Taylor, "Toward Hegemony? Assessing China's Asian Ambitions", *Security Challenges*, Vol. 1, No. 1 (November 2005), pp. 1–5.
41 Sophal Ear, "Sowing and Sewing Growth: The Political Economy of Rice and Garments in Cambodia", Stanford Center for International Development (SCID), Working Paper, no. 384 (Stanford, CA: SCID, April 2009).

42 IMF, "Statement at the Conclusion of an IMF Staff Mission to Cambodia", press release, no. 09/325, 24 September 2009.

43 In recent years, China has begun to take a less confrontational, more sophisticated, more confident and, at times, more constructive approach toward regional affairs. For more information, see: Evan S. Medeiros and M. Taylor Fravel, "China's New Diplomacy", *Foreign Affairs* (November/December 2003).

44 "China's Paths to Hegemony", *The Washington Times*, 17 January 2008, available at www.washingtontimes.com/news/2008/jan/17/chinas-paths-to-hegemony/ (accessed 15 July 2012).

45 Joshua Kurlantzick, *Charm Offensive: How China's Soft Power is Transforming the World* (New Haven, CT: Yale University Press, 2007).

46 David Shambaugh, *Power Shift: China and Asia's New Dynamic* (Berkeley, CA: University of California Press, 2005), pp. 187–203.

47 Oliver Hensengerth, "Money and Security: China's Strategic Interests in the Mekong River Basin", Briefing Paper, ASP BP 2009/02 (London: Chatham House, June 2009), p. 6.

48 Ibid.

49 "Cambodian-Chinese Friendship Benefits Two Peoples: Hun Sen", *Xinhua News Agency*, 13 September 2009.

50 Sophal Ear, "Governance and Economic Performance: Credibility, Political Will, and Reform", *Cambodian Economic Review*, Vol. 1, No. 1 (May 2005), pp. 17–49; Julio Jeldres, "China-Cambodia: More Than Just Friends?" *Asia Times Online (Hong Kong)*, 16 September 2003.

51 Ian Jeffries, *Political Developments in Contemporary China: A Guide* (London: Taylor & Francis), p. 378.

52 Brian Shen, "New Opportunity for China-ASEAN Trade", *Beijing Review*, Vol. 46 (May 2003), p. 18.

53 Paul Marks, "China's Cambodia Strategy", *Parameters*, Vol. 30, No. 3 (Autumn 2000), pp. 92–108; and for related arguments, see: Sophie Richardson, *China, Cambodia, and the Five Principles of Peaceful Co-existence* (New York: Columbia University Press, 2009).

54 Thomas Lum, *Cambodia: Background and U.S. Relations* (Washington, DC: Report, April 2009), p. 13.

55 International Institute for Environment and Development, FAO and International Fund for Agricultural Development, *Land Grab or Development Opportunity?* (London: Russell Press, 2009), p. 120.

56 Bill McKibben, *Deep Economy: Economics as if the World Mattered* (Oxford: Oneworld Publications, 2007), p. 177.

57 Steve E. Johnson, "China Emerges as World's Fastest Growing Market for Tropical Timber Products", in *Trends in the Tropical Timber Trade* (UNECE/FAO Forest Products Annual Market Review, 2001–02), available at www.unece.org/fileadmin/DAM/timber/docs/rev-02/chap-13.pdf (accessed 15 July 2012).

58 Jianguo Liu and Jared Diamond, "China's Environment in a Globalizing World", *Nature*, Vol. 435 (June 30, 2005), pp. 1–2.

59 E.S. Muskie, *Exploring Cambodia: Issues and Reality in a Time of Transition* (Center for National Policy, 1990).

60 Long Kosal, "Sino-Cambodia Relations", Cambodian Institute for Cooperation and Peace (CICP), Working Paper, no. 28 (Phnom Penh: CICP, 2009).

61 Transparency International's Corruption Perception Index 2011 ranked Cambodia 164 out of 183 countries and territories. See cpi.transparency.org/cpi2011/results/ (accessed 20 July 2012).

62 John A. Hall, "Human Rights and the Garment Industry in Contemporary Cambodia", *Stanford Journal of International Law*, Vol. 36, No. 1 (Winter 2000), pp. 119–74.

63 Sandra Polaski, "Combining Global and Local Forces: The Case of Labor Rights in Cambodia", *World Development*, Vol. 34, No. 5 (May 2006), pp. 919–32; "Worker Rights and Global Trade: The U.S.-Cambodia Bilateral Textile Trade Agreement", *Harvard Business School Case Studies* (Cambridge, MA: Harvard Business School, 2004), pp. 1–32.

64 World Bank, "Battling the Forces of Global Recession", East Asia and Pacific Update (Washington, DC: World Bank, April 2009), p. 18.

65 See: "Cambodia, Kuwait, and Farmland – Petrodollars v Smallholders", *The Economist*, 23 April 2009.

66 FAO, "From Land Grab to Win-Win: Seizing the Opportunities of International Investments in Agriculture", Economic and Social Perspectives (Rome), Policy Brief, no. 4, 2009.

67 "Anthropogenic effects are those that are derived from human activities, as opposed to those occurring in natural environments without human influences". See: European Environment Agency, *Anthropogenic Processes*, n.d., available at www.eea.europa.eu/themes/water/wise-help-centre/glossary-definitions/anthropogenic-processes (accessed 15 July 2012).

68 See: "Country for Sale, Welcome to Cambodia: How Cambodia's Elite has Captured the Country's Extractive Companies", Global Witness, February 2009, p. 6, available at www.globalwitness.org/sites/default/files/import/country_for_sale_high_res_english.pdf (accessed 15 July 2012).

69 See: Chakrya and Strangio, "Villagers Gather to Express Dam Concerns".

70 In March 2009, the Economist Intelligence Unit (EIU) ranked Cambodia fourth out of 165 countries in terms of political instability. EIU, "Manning the Barricades: Who's at Risk as Deepening Economic Distress Foments Social Unrest", Special Report, London: The Economist Intelligence Unit, March 2009.

71 WWF, *Mekong Dolphins on the Brink of Extinction* (Washington, DC: WWF, June 2009), p. 52; also see: Huma Sheikh (ed.), "Mekong River Dolphins Close to Extinction", *Beijing (Xinhuanet)*, 18 June 2009, available at news.xinhuanet.com/english/2009-06/18/content_11561164.htm (accessed 15 July 2012).

72 UN Forum on Forests, *Mekong River Commission (MRC), Mandate*, n.d., available at www.fao.org/forestry/directory/1856/en/ (accessed 15 July 2012).

73 MRC, *Meeting the Needs, Keeping the Balance: Strategic Plan 2006–10* (Vientiane: MRC, December 2006).

74 S. Burgos and M.J. Otte, "Poultry, HPAI, and Livelihoods in Cambodia – A Review", FAO, Mekong Team Working Paper, no. 3 (Rome), 2008, pp. 1–5.

75 See: Ear and Burgos Cáceres, "Livelihoods and Highly Pathogenic Avian Influenza in Cambodia".

76 C. Shay, "Scientists Announce Trove of Fragile New Species in Mekong", *Time*, 29 September 2009.

77 Elizabeth Economy, *The River Runs Black: The Environmental Challenges to China's Future* (Ithaca, NY: Cornell University Press, 2004), p. 122.

78 Carl Middleton, *Cambodia's Hydropower Development and China's Involvement* (Berkeley: International Rivers Report, January 2008), pp. 1–77.

79 M. Babel and S. Wahid, *Freshwater Under Threat – Southeast Asia* (UNEP-AIT Report, 2009), pp. 1–44.

80 Elizabeth Economy, "China's Rise in Southeast Asia: Implications for Japan and the United States", *The Asia-Pacific Journal – Pacific Focus*, 6 October 2005, available at www.japanfocus.org/-Elizabeth-Economy/1815 (accessed 15 July 2012).

81 Françoise Nicolas, "Chinese Direct Investments in Europe: Facts and Fallacies", Briefing Paper (London: Chatham House, June 2009).

82 Sophal Ear, "Cambodian 'Justice'", *Wall Street Journal Asia (Hong Kong)*, 1 September 2009.

83 Michael Glosny, *Meeting the Development Challenge in the 21st Century: American and Chinese Perspectives on Foreign Aid* (Washington, DC: National Committee on US-China Relations, 2007).
84 Sophal Ear, "The Political Economy of Aid and Governance in Cambodia", *Asian Journal of Political Science*, Vol. 15, No. 1 (April 2007), pp. 68–96.
85 Sophal Ear, "Governance and Economic Performance: Credibility, Political Will, and Reform", *Cambodian Economic Review*, Vol. 1, No. 1 (May 2005), pp. 17–49.
86 Avery Goldstein, "Parsing China's Rise: International Circumstances and National Attributes", in Robert S. Ross and Zhu Feng (eds), *China's Ascent: Power, Security, and the Future of International Politics* (Ithaca, NY: Cornell University Press, 2008), p. 55.
87 Editorial, "Cambodia's Oil Resources – Blessing or Curse?", *The Economist*, 26 February 2009, available at www.economist.com/node/13184945 (accessed 15 July 2012).

III
Conclusion

6 Assessing opportunities, mitigating risks

Part I presented an introduction to China, its growth, its energy needs and its relationship with the world. Part II presented three comparative case studies to examine China's actions and initiatives in Angola (Africa), Brazil (Latin America), and Cambodia (Asia). Now, this last part brings all the information together to compare and contrast issues and options confronted by Beijing in terms of vital interests, national security, economic progress and potential pitfalls. Part III has two chapters. The first deals with assessing opportunities and mitigating risks. The second concerns China's move from strategies to actions, along with concluding remarks. Because this section presents new information, it should not be entirely considered as conclusive, but more in the light of connecting the dots as critical factors become clearer to us.

At the beginning, there was need

Overall, it is quite clear that world oil consumption shows few signs of declining in the 21st century. Diplomats and politicians promoting *radical* shifts toward increased investments in sustainable alternatives and renewable energies sources, as well as lifestyles based on ecological sustainability, have *not* signalled the culmination of dependence on coal, gas, oil and its derivatives. Industrialized nations, led by Canada, European nations, Japan and the USA continue to demand and consume high volumes of coal, gas and oil to fuel consumer needs, national industries,[1] airlines, commercial sea freights, primary and secondary infrastructures, and the military (i.e. jets, helicopters, trucks, ships, etc.). To the list of industrialized nations we need to add the BRIC economies (Brazil, Russia, India and China), which are, to be sure, picking up speed as their economic engines compete on an equal footing with old powers, but also redefine the way energy is used by their people.

As a result of the rapid gas and oil consumption at the hands of old and new powers, the global supply of energy, materials and resources has been strained.[2] Copper, gas and oil are more expensive, some minerals are harder to find, crude deposits are difficult to discover, and deforestation in key ecosystems is reducing the availability of precious and non-precious woods needed for furniture making, box and pallet making, and construction of buildings and homes.

As noted earlier, oil has been a sought-after commodity that shows no signs of cooling off any time soon. "World crude oil and liquid fuels consumption will grow from its record-high level of 87.1 million barrels per day (BPD) in 2010 to 88.2 million BPD in 2011, and will reach 89.6 million BPD in 2012".[3] Consequently, many Organization of the Petroleum Exporting Countries (OPEC) and non-OPEC countries have seen an explosion of exploration and drilling agreements. These accords, contracts, concessions, deals, arrangements and pacts have been tabled, negotiated and signed at unprecedented rates in the last decade and a half of this young century. While it is true that the deal-making machinery has slowed owing to the global financial crisis, it has become readily evident that one country has taken advantage of its unscathed position to emerge as a creditor of last resort to many oil- and non-oil-producing states. This country is China.

The Chinese, sitting on trillions of US dollars in foreign exchange reserves,[4] have extended credit lines, grants and loans to countries in need of cash injections in exchange for: 1) principal plus interest; 2) access to gas and oil; 3) payment with gas and oil; and 4) payment with minerals or other critical inputs. This business-banking strategy is aligned with national plans and Chinese Communist Party (CCP) ambitions of pushing the country towards greater urbanization (i.e. a more modern society with developed infrastructure all around), developing a booming consumer auto industry counting on the fact that millions of Chinese will soon be able to afford low-cost transportation, propelling the construction of mega real estate projects as well as engineering and industrial projects, and to continue commanding the position of the manufacturing hub of the world.[5] However, as easily deduced, all these plans, dreams, ambitions and objectives demand increasing amounts of energy sources, raw materials and natural resources.

This *need* is what drives Beijing's policies and, at the same time, it is this "go abroad" strategy that allows Chinese public and private oil enterprises to show up in foreign outposts.

Brief outline of the chapter

This chapter will undertake an assessment of opportunities and of risk mitigation. The following sections will look at China's entry points for insertion as the world engine of growth and as a stabilizer of regional and global financial systems. It starts from the premise that coal, gas and oil were pivotal to China becoming a manufacturing hub for the world to enjoy cheap trinkets, clothes, shoes and household products, and that today gas and oil are integral to China's industrialization process so that it can fully insert itself as an equal player with the old powers. China's thirst for gas, oil and minerals is so evident and notorious that, at a glance, it is highlighted as the first area to *strangle* the country out of prosperity. Evidently, Beijing knows this too. It knows that risks and threats are plentiful and that others know it too (i.e. the risk of interdiction at the Strait of Malacca and Hormuz). As a recent player and nascent economy, it has not the experience old powers have earned over many years. In response, China has laid out a diversified portfolio of foreign gas and oil and mineral deals to secure its national interests. The emerging asymmetries that may impede Beijing's progress are analysed.

Assessing opportunities

The advent of the 21st century marks an era of tremendous opportunities for China and its people. China is no longer thought of as the backward country that it was once considered. The new China is more culturally aware, dynamic, constantly morphing, more open to the world, vibrant and increasingly wealthy. This change is no mere happenstance; China already had a plan in place. Following a period of power consolidation, Chairman "Mao launched the First Five-Year Plan from 1953 to 1958. The plan aimed to end Chinese dependence upon agriculture in order to become a world power".[6] It is here, at the beginning of the 1950s, that Chinese leaders understood the slow progression from a primary-dependent to a tertiary-led economy was the road to success and the surest route out of poverty. After enjoying "Soviet Union's assistance for some years, new industrial plants were built independently and agricultural production eventually fell to a point where industry began to produce enough capital that China no longer needed Russian support". It was the sudden and short-lived success of the First Five-Year Plan that encouraged Chairman Mao to instigate the Second Five-Year Plan. This was to be called, in accordance with the dreams of this boss, the Great Leap Forward, in 1958–59. At the same time, Mao launched a phase of rapid collectivization. This second plan did not turn out to be successful: "nation-wide political campaigns led by Mao, such as the Great Leap Forward and the Cultural Revolution, are considered catastrophic failures"[7] that caused millions of deaths. To this day, "Chairman Mao remains a controversial figure, with a contentious legacy that is subject to continuing revision and fierce debate"[8] among CCP leadership, historians, and national and foreign scholars.

It was not until much later that China enjoyed the fruits of market openness and liberalism. The rise to power of Deng Xiaoping in 1978 brought a wave of change that put China on the right track after Mao's monumental failures (and failings). "As leader of the Chinese Communist Party, Deng was a reformer who led China towards a market economy". Indeed, "the domestic social, political and, most notably, economic systems would undergo significant changes during Deng's time as leader". The goals and objectives of "Deng's reforms were summed up by the Four Modernizations, that is, those of agriculture, industry, science and technology, and the military". Certainly, "Deng's strategy for achieving these aims of becoming a modern, industrial nation was the *socialist market economy*",[9] which is now more commonly known as autocratic capitalism.

Since then, China has experienced rapid economic growth. From the late 1970s until the early 1990s, "China decided to accelerate its modernization process by stepping up the volume of foreign trade, especially the purchase of machinery from Japan" and Western countries. "By participating in narrow export-led growth, China was able to step up the Four Modernizations by attaining foreign" capital, access to markets and "advanced technologies, and management experience, thus accelerating its economic development".[10] Since the 1990s, much of China's growth has stemmed from prudent monetary policy and the opening of its borders to foreign direct investment, among other things. As evidence of the success of these decisions, "from 1978 to 2004, China's GDP [gross domestic product] grew by an annual average of 9.5 percent".[11]

Based on its success, China has managed to grow successfully under an alternative system. In previous chapters we learned that this is now called the "China Model". This model is nothing more than the *coexistence* of economic freedom and political repression. For countries that do not reject this idea, it is a model for possible emulation. As it is, Cambodia, Myanmar (Burma), Venezuela and Vietnam are on board—or at least seem to be, when one listens to their rhetoric.

Clearly, China's move from a centrally planned economy to an export-led, market-oriented economy embracing a number of Western features marks a turning point in the country's relationship to businesses, markets, clients, resources and the world at large. Also, it signals the beginning of its dependency on foreign gas and oil, and its massive use of coal-based growth. This dependency arises because the manufacturing and industrialization boom is impossible without advanced energy infrastructure, knowledge, drilling equipment, scientific backing and expertise necessary in upstream (i.e. exploration and drilling) and downstream (i.e. refinement and sale of the oil) operations. However, this is only a temporary setback. Chinese officials have a plan and they formulated the details during the process of execution. Currently, China counts with major national oil companies (NOCs),[12] which manage, own, or operate some of the most abundant and profitable oil fields in the world.

An opportunity presents itself to China to manoeuvre into the energy business, an industry that for a long time had been dominated by OPEC and old powers. As noted earlier, much of China's foreign oil was obtained from upstream assets. Now, though, China secures crude oil "through downstream deals, rather than by buying directly into upstream assets. Such deals serve to secure crude supply at a much faster rate than direct upstream acquisition, although the crude is delivered from a foreign location. It also serves to redress China's growing shortage of refinery capacity at home". At this point, it is "an open question as to how far China will want to see its oil dependency go, but China's foreign oil dependency will soon equal that of the United States, which some … [US] … observers have suggested is uncomfortably high. China's thirst for hydrocarbons led it to abandon the ideal of oil self-sufficiency, back in the late 1980s".[13]

Because millions of urban and rural Chinese have been brought out of abject poverty through economic and state modernization, China's leaders have an opportunity to ask the Chinese people to stave off recurrent criticisms levelled by internal and external constituencies on environmental degradation, water and air pollution, lack of civil freedoms, human rights violations and erratic labour standards, among many others. Mostly, China's leaders have won that bargain, but increasingly Chinese people have become vocal, if not restive, in recent years. The potential is there for China to continue proving that its iron fist over its people is legitimate while at the same time proving to the world that alternative growth models are, indeed, viable.

Another opportunity that emerges is one for domestic consumer-oriented industries. This is because the Chinese workforce continues to earn higher wages, and this propels increased consumer spending. One industry that will gain disproportionately from this massive consumption boom is the automobile industry. As urban growth moves up, so does the need for vehicular transportation (including motorcycles). Tellingly, "after China's entry into the World Trade Organization in 2001, the development of the automobile market further accelerated. Between 2002 and 2007, China's national automobile market grew by an average 21 percent, or one million vehicles year-on-year".[14] For instance, in 2010 China became the world's largest car market in terms of sales volume. Recently, "data released by China Association of Automobile Manufacturers" for July 2011 shows "that automakers in China sold 9,325,200 new vehicles in the first six months of that year, 3.35 percent more than a year ago".[15] "The number of registered cars, buses, vans, and trucks on the road in China reached 62 million in 2009, and is expected to exceed 200 million by 2020".[16] All these commercial vehicle sales are occurring at an increasingly rapid pace despite the central government's efforts to deter car sales in Beijing, Shanghai and other major cities to curb traffic congestion and air pollution.

In terms of industrial output, oil will continue to be a major driver of industrial processes in China. "Industrial value-added output measures the final results of industrial production. In other words, it is the value of gross industrial output minus intermediate inputs (such as raw materials and labour costs)".[17] Industrial output has followed an upward trend in the last decade. For example, according to the Chinese National Bureau of Statistics, "China's industrial value-added output grew 14.3 percent year-on-year in the first half of 2011".[18] This is great news for industrialists and also for the economy as a whole. It is exactly here that CCP officials have detected a number of opportunities for investments in what is sure to be the wave of the future. In the clean energy race that has started after the climate crisis put everyone on alert, "China is quickly acting on its strategy to dominate the solar industry".

How is China doing this? Simply, "by pushing down solar panel prices and building assembly plants on U.S. soil". In reaction, Western analysts fear that "American solar manufacturers may not be able to compete for long".[19] To this end, "China has developed a straight-forward, no-nonsense approach to achieving its 2 gigawatt solar capacity target by late 2011, and gaining leadership in the solar industry". Its goal is to "build market share. With the help of government investment", incentives, scientific breakthrough, and other forms of stimulus, "China is on the path to achieving this lofty goal". For example, "Chinese companies like Suntech Power Holdings have already succeeded in driving solar panel price reductions. Furthermore, China is circumventing protectionist legislation by constructing assembly plants in the U.S." The opportunity lies in Beijing offering solar companies the best in financial assistance possible in the form of state subsidies, free land, money for research and development, and extremely low interest loans (just as it does with oil ventures abroad). With so many incentives in place, Chinese entrepreneurs will be "more than willing to enter the solar market and to capitalize on the growing industry".[20]

As things stand in the world with Europe on the brink of collapse and the USA in political deadlock, China may very well emerge as a more balanced and stable government than the democracy- and freedom-loving countries in Europe and in the USA: they have much to win from this.

Mitigating risks

It is not surprising that "there is already high risk that Beijing will produce unpleasant foreign policy surprises" in the year of writing this book, many of which are related to policy differences with the West. Much of this risk stems from rising nationalism, "its ongoing political transition, and the leadership's unwillingness (and perhaps inability) to resolve internal debates about China's role in the world". Retired military officials and former top diplomats appearing on the nightly news are quick to point out that "Beijing will be more apt to meet provocation with provocation in months to come, using both its naval and economic powers. A harsh Chinese response to an incident at" the South China Sea involving a US ship or an "American ally would provide a significant test"[21] for the party that wins the US elections in November 2012. Clearly, "America's military presence in the Asia Pacific has rarely been contested, even by the Chinese".[22] However, this does not mean that it cannot happen. As far as Beijing is concerned, CCP officials will continue to abide by their mottos of peaceful rise, benign collaborations, embracing of international law, and growth.

A case in point is its bittersweet relationship with North Korea. To China, having a hot-headed neighbour that stands its ground in not allowing old powers to meddle in domestic affairs is not ambiguous, insulting or odd, for it invalidates some of the notions of non-intervention that Beijing preaches all around the world. It is clear, however, that China has to proceed carefully with North Korea, as it is prone to provoke South Korea and Japan with its military exercises, rocket launches, incendiary exhortations and military parades. It is not unreasonable for North Korea to remain belligerent and dangerous if it feels threatened and pushed, and more likely to become a law-abiding nation if it is approached and engaged collaboratively and respectfully.

As with the USA, the relationship between China and South Korea has been more a marriage of practical convenience. South Korea is intent on asserting a national identity that is separate from China, the USA and other Asian countries, and it is focused on promoting peace, generating prosperity for its citizenry, and working with allies to tame the belligerent ways of its northern neighbour. While both China and South Korea still value their strategic alliance with Washington, they do not wish to be belittled, patronized or talked down to. This is key for CCP officials to understand; they should not mimic the hegemonic acts of the USA in its backyard. Ultimately, however, China's need to keep North Korea as a buffer between itself and South Korea (or what could be a unified Korea in which the South absorbs the North) is paramount. This explains why the arrest and detention of 29 Chinese fishermen by North Korea in May 2012 for almost two weeks did not perturb China's relationship with North Korea. The fishermen were allegedly beaten, robbed and stripped, and allowed to eat only starvation rations by their North Korean captors (who acted more like pirates than soldiers).

In a wide sense, to mitigate risks China is well advised to engage in constructive US-Chinese relations in areas that already have their share of tensions and that can be easily eased via policy instruments and tangible initiatives. For example, some of the areas that both countries could improve upon are cyber-security issues and copyright enforcements (to promote indigenous innovation). These problems, if not addressed, "will make regional tensions more difficult to manage". Evidently, nobody in Asia "wants a security confrontation that would undermine economic growth" and the progress of the ongoing recovery, but the enhanced US security presence in Asia and the Pacific does embolden "China's neighbours to take on more assertive" and bold "policy positions with" Beijing, "especially on strategic issues"[23] that are very important to them.

Another area of work is on public image. For example, Beijing has a long road ahead to clarify that China's success is not merely due to its plentiful supply of low-wage labour. Today, China produces an increasing number of academics, engineers, Olympians, researchers and scientists. In 2003 it became the third country to launch a man into space, which has long been an act that qualifies a nation as a superpower. Also, its research and development budget grows year by year; it experiments with new designs, systems, techniques and procedures in such fields as bioengineering, biotech, civilian nuclear power, environmental technologies, pharmaceuticals and medical advancements. Chinese universities and research centres could earn goodwill and respect from other countries in the world by opening themselves to more collaboration, student exchanges, cultural reciprocity via embassies and consulates, and for-profit cooperation in areas that for too long have been restricted to others, due to national security issues and copyright infringement suspicions. Despite the wonderful infrastructure and amazing laboratories in universities, who, as a graduate student, would want to do research in a country where the Internet is censored heavily and the use of a virtual private network (to make a tunnel through the Great Firewall) is a prosecutable offence?

China could also mitigate risks by downplaying disputes and contestations in the South China Sea. Yes, it is well known now (and Beijing has been good at reminding its neighbours) that China is forcefully asserting its exploration rights and claims of sovereignty over this body of water.[24] Incidents in the South China Sea intensify longstanding international questions over the strategically critical, potentially energy-rich sea that have become more urgent in 2012 (the year of writing this book) as the long-dominant USA and fast-growing China both seek to increase their naval power in the region.[25] "Countries such as Vietnam and the Philippines—both entangled in boundary disputes with Beijing in the South China Sea—could decide that closer defence and security ties with Washington provide the cover needed to push back against perceived Chinese advances" in the region. A wrangle "could create a maritime confrontation that might draw in the U.S." A miscalculation of such proportions could "raise a host of risks and threats for the Asian security"[26] environment and for global markets in 2013, all of which are undesirable scenarios that neither China nor the USA is willing to even consider at this point.[27]

China acknowledges that its economic transformation has altered its geopolitical landscape and its policy options. This is a recognition that has been closely followed by all countries in the international system, especially since everyone else is competent enough to realize that economic growth requires resources, and the Chinese have been seen prowling international markets for gas, oil, minerals and timber to sustain the speed of this growth. The West has kept close tabs on Beijing's relentless work to establish ever-new supply relationships on every continent. Of relevance, too, is the fact that among its partners are the governments of Angola, Iran, Myanmar, Sudan and Venezuela—not exactly the hallmarks of liberal democracies. Here is another area where Beijing could mitigate risks: sharing with the rest of the world some of the most fundamental ideals, norms, principles and values that the majority of countries embrace so that China is not perceived as an outlier, but as a team player. Of course, China's nickname of "Gang of One" bodes ill for this prospect.

Another avenue to reduce its exposure to criticism and to prove its willingness to play by rules set in the *region* is for Beijing to continue to organize regional meetings from which Brussels and Washington are excluded: East Asian summits, conferences, congresses and symposia, as well as participation in cooperation and development organizations. It has already done this to highlight its assertiveness and independence, but it must now be tailored to really demonstrate that it is willing to function as a *regional leader* that can assume a number of the roles that the British and the Americans once assumed in the region, but this time focusing on the positives, while underplaying the usual suspicions of neighbours that China is out to control them.[28]

Certainly, in Southeast Asia defence and security misgivings begin with the fact that the gap in international stature that has existed between the USA and China has narrowed consistently over the last two decades. In the years after World War II, most Asian nations felt closer to a benevolent and powerful Washington than to an impoverished and turbulent Beijing. Today, however, China is associated with vibrant dynamism and economic success, to the point that this has been attracting a growing number of immigrants who see China as the "next" land of opportunity—much as the USA was perceived back in the 1960s and 1970s. The above highlights the "realities of Asia today. For many countries in the region, China's economic development is a source of lucrative new business opportunities, but they want to avoid becoming too economically or politically dependent on Beijing". The fear is that CCP officials in Beijing might turn out to use this dependency as a tool for abuse, exploitation, manipulation or subjugation.

The Chinese system of authoritarian growth (as opposed to democratic growth) is admired by leaders who want their countries to find a niche in the world economy, but who do not want to risk losing their own power. This model of growth is supported by Chinese financing, which helps governments in Africa, for example, to undertake capital projects without the kind of accountability and transparency that is required by the World Bank. Beijing is at a position where it could act as an agent of positive change by helping countries move towards prosperity and an ensuing blanket of benefits that touches as many social classes as possible.

Although it is a problem everywhere in the world, the widening income gap between rich and poor in China clashes acutely with the country's socialist pedigree. That pedigree emphasized egalitarianism. The often-quoted phrase that "inequality breeds resentment" is especially true in China, where people suspect that the rich obtained their wealth not through hard work, but through corruption and cronyism. The Chinese are also angry about pollution, bitter about industrial projects that drive them off their land, and worried about uncontrolled disease outbreaks. It is here, along the lines of these issues, that CCP officials have much work to do in assuring its citizenry that China's wealth will not be concentrated in the top 10% as in the West, but that the Government will work hard to make sure a balanced and fair distribution of wealth will take place in their lifetime.

Some reflections

It is now clear to the international community that Beijing has an arduous task in protecting its national security interests. Gas and oil security is the most important part of achieving energy security.[29] For instance, according to the latest *World Energy Outlook*, "global gas demand will rise by more than 50 percent over the next 25 years, as gas outstrips coal to come close to equalling oil in the energy combination. Meeting that demand will require a sharp increase in production, equivalent to three times the amount of gas produced by Russia today, which the International Energy Agency estimates will be met by a mixture of conventional gas and shale-gas, as well as some other unconventional forms of this fuel, such as coal-bed methane". As expected, "China could easily become both a principal producer (its shale-gas resources are estimated to be the largest in the world) and perhaps the largest importer".[30]

Without energy security China could have difficulties in maintaining its growth rates to stave off dissatisfaction, mass demonstrations and unemployment. After all, those Chinese officials who were educated in the West have learned that energy security is one of the major principles followed by old powers. However, like most high-polluting countries, China is now looking at *alternative energy sources* to supplement its growing dependence on foreign oil and domestic coal. There are a number of opportunities in this area, but they will come at a cost to the Chinese Government. There is very little reason to believe that China's effort to attain energy security will fail given that the private sector waits on the sidelines for a stable market to develop and for technological advances to move forward in order to throw themselves behind clean energy alternatives. Meanwhile, however, economies of scale do not warrant a significant shift away from coal, gas and oil. The reality for China is that its own domestic capacity to produce oil is low and alternative energy sources are still too expensive or still too inefficient to eclipse gas and oil as a key commodity in pushing the rapid industrialization plans of the country. Indeed, China is not positioning itself to be totally reliant on other renewable sources of energy in the long term. However, it does want to be less dependent on foreign oil owing to the geopolitical and geostrategic security threats it faces, coupled with ever-increasing environmental concerns.

Chinese oil companies invest in, and deal with, Iran, Sudan and Venezuela, as well as some key suppliers of old powers; these alliances and partnerships are perceived as potential threats to US national security and trade imperatives. The potential for retaliation or restrictive measures against China should there be problems or friction or contention between parties is a source of risk for China, as it would elicit a series of counter-responses. This is a risk that China must mitigate; hence, CCP officials must tread carefully in these areas. Already, some of these alliances and partnerships have unravelled diplomatic efforts aimed at thwarting these antagonistic states, with China stuck in the middle of it, too.

Moreover, the relationships are not limited to oil and by-products. Chinese companies provide Iran with informational, computing and engineering technologies to aid Iranian industrial, mining and energy industries. This runs counter to UN "sanctions against Iran for its nuclear"[31] proliferation activities. Separately, China's oil consumption fills Angolan and Sudanese coffers with revenues that finance weapons that have been used to infringe the sovereignty of neighbours. There is a risk that the USA could stiffen its diplomatic position with China. Sanctions and threats may continue to be ignored by China as it continues to deal with its rogue oil partners, but at some point it is believed that the West will react and may even do so pre-emptively. In reality, despite commentaries of CCP belligerence, the Beijing-Washington relationship is based on a wary balancing of interests, a tip-toeing act that is not too small on assurances and not too big on sensitivities.

On a similar footing, sensitive negotiations and tip-toeing have been undertaken between China and the Saudis. At this point, China is Saudi Arabia's largest energy trading partner. At the same time, China happens also to be the largest foreign investor in Iraq's oil fields as well as a key energy importer of oil and by-products from Iran. At a glance it is easy to determine that Beijing has emerged as an important player in a region the preserve of which was limited to old powers. The arrival of new powers to the scene is worrisome to some states undergoing revisionist policies. Important, too, is the fact that the world's largest oil reserves are centred in the Middle East, and China has emerged as this region's most cherished partner, to the point of sidelining Europe and the USA. We believe that the geopolitical and military environment in this strategic region will be disturbed by China's presence, and that the European Union and the USA will react when either senses intrusion in a region where both have enjoyed unlimited access for decades.

What does China's meddling in the oil business means for the USA? Assuredly, at a time when the US military is stretched thin across the globe, with a still ailing domestic economy recovering from the economic downturn of its own making, it must continue to monitor Chinese oil investments abroad in order to follow-up on areas that may present critical threats to vital security interests. It would be speculative to say what reactions may be engaged by the US Government, but the Chinese insist that they do not pose egregious geopolitical threats to the USA in the short term and that they play lawfully and by the rules of engagement. Through history, the rise of a new power has often led to war or conflict, whether caused by the desire of that power to spread its wings or the attempt by rivals to clip those wings. China's rise, while possibly a win-win for all involved (especially if the decline of the USA is not absolute but only relative), is likely to lead to contention or violence.

Taiwan: the ultimate quagmire

Taiwan was part of China from the mid-17th century until the end of the 19th century.[32] Then, China lost Taiwan—the fact that the island was captured by the Japanese only deepens China's determination to recover it, even though the Japanese are long gone. Chinese military officers are taught that returning Taiwan to the motherland is the last unfulfilled aspect of their careers. In 2011 President Hu Jintao sought to move beyond the threatening rhetoric that long characterized Beijing's response to Taiwan's refusal to unify with the mainland. His government has talked of ending the state of hostility with Taiwan. He is quoted as saying that "achieving reunification by peaceful means best serves the fundamental interests of all Chinese, including our Taiwan compatriots … heal wounds of the past and work together to achieve the great rejuvenation of the Chinese nation". Above all, China regards Taiwan as a part of its territory and has threatened to invade, should Taiwan seek formal independence; however, Beijing has flexed its diplomatic arm by making appeals "to further relations with Taiwan, saying they should move beyond the history that divides them and *focus* on common economic and cultural interests".[33]

Over the years, Taiwan has strengthened its relations with the USA, to the point that many exchanges between the countries have been loaded with assurances of protection in the event of aggression by Beijing. However, it must be stressed that Washington is not formally committed to rescuing Taiwan in the event of an attack directed by Beijing because Americans do not want the Taiwanese to be so sure of protection that they go provoking Beijing. Similarly, the USA does not want the Chinese to think that they can attack Taiwan and get away with it, without recrimination.

The ultimate rationale for keeping American military forces deployed in the Asia-Pacific region is to deter potential aggression initiated by anyone. This is all the more necessary once the USA is aware of the domestic pressures that could drive Beijing to behave belligerently. If a crisis arises, the most possible scenario laid out by Western powers is that Chinese decision makers will look out to the Pacific Ocean and be confronted by a mighty American naval force with the ability, capacity and determination to defend its strategic national interests, but also to resolutely defend Japan, Taiwan, South Korea and other Asian allies and friends.

Former US Secretary of State Madeleine Albright said that East Asia will be stable, provided one country does not try to bully or intimidate another into doing something it does not want to do. As of now, neither China nor Japan (especially with its more than 230% debt-to-GDP ratio) seem possessed of imperial ambitions in the region, yet neither will appreciate being seen as the junior partner or the lesser state. South Korea wants to focus on growth, development and peace, whereas North Korea looks for international recognition and for a way to perpetuate its dreams of non-abidance and self-reliance. Taiwan will continue to face a choice between provocation and restraint.[34]

On a more pragmatic view, Beijing is pushing a plethora of free-trade deals with its neighbours, with the hope that these actions might drive Taiwan closer to China. Taiwanese officials, for their part, are "willing to take the political risk of tying a self-ruled democratic island economically to its giant authoritarian neighbour because of the rest of the world's craze for free-trade deals" and the prospect of benefiting from China's economic rise in the region. Currently, "Taiwan's exports to China face tariffs ranging from 5 to 15 percent and its government fears that, unless these tariffs are lowered, the island will be left at a competitive disadvantage in the giant Chinese market". However, "many Taiwanese, including the pro-independence opposition party, fear that the proposed" accords and deals on free trade and market openness are "really a ploy by" Chinese officials "to bring about unification by stealth".[35]

It must not be forgotten that Chinese foreign policy is based on the principles of sovereignty and non-intervention in the internal affairs of others.

This being the case, an outright attack on Taiwan or direct intervention in the dealings of other nation-states might be quickly interpreted as a violation of this widely shared promise. Also, it might validate the suspicions in Southeast Asia that Beijing secretly covets control of the region. This outright inveiglement of intentions is not welcomed by the West, particularly when it has to do with the economy and military build-ups. For example, "China won't say how it invests its foreign-exchange reserves, which have grown rapidly over the past decade. Beijing has used its control over the exchange rate as a key plank of its economic-development strategy and has racked up immense trade surpluses".[36] Just how much European and US debt is possessed by China is a matter of speculation, though it is believed that China owns more than $1 trillion of US Treasury bills. Still, a source of pride for the CCP is its advancement of the economic prospects of its people, considered *more* important by many who are destitute and poor than the right to vote or speak freely. In Beijing, the lesson learned from a volatile world is clear: stability is the ultimate goal. It is for this reason that dissident movements are crushed in China, religious liberties circumscribed, artists critical to the government are imprisoned, political organization is prohibited, the Internet is monitored and censored, and foreign criticism is not allowed to jeopardize internal order. The CCP has realized that the decline in the US's world-wide reputation has reduced pressure on China to improve its dubious record.

Notes

1 In the first six months of 2012, in the USA both Democrats and Republicans bickered over the feasibility and viability of approving "the Keystone Pipeline System, a pipeline system to transport synthetic crude oil and diluted bitumen from the Athabasca Oil Sands in northeastern Alberta in Canada to multiple destinations in the U.S., which include refineries in Illinois, the Cushing oil distribution hub in Oklahoma, and proposed connections to refineries along the Gulf Coast of Texas". Proponents argue that "the pipeline would allow the U.S. to increase its energy security and reduce its dependence on foreign oil", but critics note that approving it will delay much-needed efforts to move the economy towards renewable energy sources, and the concomitant job creation that come with it. For more information, see: en.wikipedia.org/wiki/Keystone_Pipeline (accessed 15 July 2012).

2 "Oil's share of primary energy has declined every year over the past decade, while coal's share of the total has increased by four percentage points since 2000. The main reason for this is China. In 2000, China consumed 11% of the world's energy; in 2010 it consumed 20.3% of a significantly bigger pie, making it the biggest energy consumer on the planet". See: "Energy Statistics: The World Gets Back to Burning", *The Economist*, 8 June 2011, available at www.economist.com/blogs/schumpeter/2011/06/energy-statistics (accessed 15 July 2012).

3 See Total Petroleum Consumption, available at www.eia.gov/cfapps/ipdbproject/IEDIndex3.cfm?tid=5&pid=5&aid=2 (accessed 15 July 2012).

4 According to the *Wall Street Journal*, "China still remains a strong buyer of U.S. debt. China's holdings of U.S. securities rose 7 percent to $1.73 trillion as of 30 June 2012, an increase of $115 billion from 12 months earlier"; see: Tom Orlik and Bob Davis, "Beijing Diversified from U.S. Dollar", *Wall Street Journal*, 2 March 2012,

available at online.wsj.com/article/SB10001424052970203753704577254794068655760.html (accessed 15 July 2012).

5 We contend that if China starts cutting back its purchases of securities around the world, national and international interest rates could climb, damaging the world economy and notching up governments' borrowing costs. If the world economy slows down, Chinese exports will surely slump, thus creating a vicious cycle of destructive concatenation—the world needs China to succeed as much as China needs the world economy to restart.

6 Famous People, *Mao Zedong*, n.d., available at www.thefamouspeople.com/profiles/mao-zedong-56.php (accessed 15 July 2012).

7 "Attitudes Towards Capital Punishment, Who Goes to the Gallows? Concern Over a High-profile Death Penalty", *The Economist*, 28 January 2012, available at www.economist.com/node/21543593/comments (accessed 16 July 2012).

8 Wikipedia, Mao Zedong, n.d., available at en.wikipedia.org/wiki/Mao_Zedong (accessed 16 July 2012).

9 Wikipedia, Deng Xiaoping, n.d., available at en.wikipedia.org/wiki/Deng_Xiaoping (accessed 16 July 2012).

10 Ibid. As noted in Chapter 1 and Chapter 2, since free trade and open markets took hold in China, their economy has grown at a sustained rate unequalled by any country in history. Extreme poverty has declined sharply, the middle class has broadened, and Chinese industrial centres have become irresistible to foreign investors, especially US companies.

11 Jiyao Bi, "China's New Concept for Development, China in a Globalizing World", UNCTAD, pp. 105–24, p. 105, available at www.rrojasdatabank.info/chinaglob05/chinaglob05p5.pdf (accessed 16 July 2012). However, forecasting growth based on past experience can also be risky. CCP officials in Beijing have "suffered a decades-long anxiety attack at the thought that the global marketplace would not be able to meet China's demand for raw materials and natural resources. But now they have exactly the opposite problem. Traders and speculators are using abandoned warehouses, parking lots, storage facilities and granaries to store all the surplus iron ore, copper, coal and other metals and minerals they have accidentally bought in preparation for a sustained economic expansion that is not happening" as fast as anticipated. See, for more information: Joseph Sternberg, "The Price of Beijing's Resources Obstinacy, Planners Used to Win Kudos for Farsightedness. But They Keep Missing the Most Important Signal of All", *The Wall Street Journal*, 6 June 2012, available at online.wsj.com/article/SB10001424052702303753904577449760270296808.html (accessed 16 July 2012).

12 These are CNOOC, CNPC, PetroChina, and Sinopec, all Chinese state-owned companies.

13 Richard Krijgsman, "China Looks to Downstream Deals as Upstream Acquisition Strategy Fails to Deliver", *Evaluate Energy*, 12 April 2012, available at www.oil-blog.com/sector/india/china-downstream-deals-upstream-acquisition-strategy-fails-deliver/ (accessed 16 July 2012).

14 Elisenda Pujol, "VW in China, EU-China Business Management Training Project", MBA manuscript, 2010, available at www.ceibs.edu/images/bmt/research/2012/02/08/4F25743E1D989EC60C4CE797F833670B.pdf (accessed 16 June 2012).

15 China Auto Web, 8 March 2012, available at chinaautoweb.com/category/sales/page/2/ (accessed 16 July 2012).

16 *Smart – Our Business*, SGS, n.d., p. 34, available at www.sgs.com/~/media/Global/Documents/Financial%20Documents/SGS_AR_2011_Book1.pdf (accessed 16 July 2012).

17 Xinhuanet News, *China's Industrial Value-added Output up 14 pct in July*, 9 August 2011, available at news.xinhuanet.com/english2010/china/2011–08/09/c_131038629.htm (accessed 16 July 2012).

18 Xinhuanet News, *China's Industrial Value-added Output up 14.3 pct in H1*, 13 July 2011, available at news.xinhuanet.com/english2010/business/2011-07/13/c_13981848.htm (accessed 16 July 2012).

19 "There is already plenty of evidence to suggest that China and other countries are beating the U.S. in solar production and R&D. For instance, California solar panel maker Solyndra and Beacon Power, a Massachusetts energy-storage firm, declared bankruptcy in 2011. Solyndra received a $528 million federal loan, while Beacon Power got a $43 million loan guarantee. Abound Solar of Loveland, Colorado, suspended operations in mid 2012. The company received roughly $70 million from the Energy Department before officials froze its credit line". See: "Colorado Solar Company Says it Will Declare Bankruptcy, Cut 125 Jobs", *Washington Post*, 28 June 2012.

20 Robert Kaapke, "The Giant Awakens to Solar Power While America Sleeps Evolution Solar Grows Roots in China", *Reuters*, 2 September 2009, available at www.reuters.com/article/2009/09/02/idUS160627+02-Sep-2009+BW20090902 (accessed 16 July 2012).

21 Ian Bremmer, "Top Risks for 2010", Eurasia Group, New York, n.d., eurasiagroup.net/pages/top-risks-2012/ (accessed 16 July 2012).

22 Yoshiko Sakurai, *Visit of JINF Delegation to Washington, D.C.*, Japan Institute for National Fundamentals, April 2009, available at en.jinf.jp/news/archives/177 (accessed 16 July 2012).

23 Ibid.

24 China is one of 162 countries that ratified the Law of the Sea treaty in 1982. The USA has not.

25 Jane Perlez, "Beijing Exhibiting New Assertiveness in South China Sea", *The New York Times*, 31 May 2012.

26 Yoshiko Sakurai, *Visit of JINF Delegation to Washington, D.C.*, Japan Institute for National Fundamentals, April 2009, available at en.jinf.jp/news/archives/177 (accessed 16 July 2012).

27 The South China Sea serves as an essential waterway for not only the Japanese and US Navy but also for a large portion of the world's trade. For instance, two-thirds of the world's natural gas trade passes through the South China Sea. The sea is the main passageway for oil from the Middle East to China, Japan, South Korea and the rest of Asia, which explains its importance. It is worth noting that Beijing argues that freedom of navigation comes into force only 200 nautical miles from a nation's coast, an argument that contravenes the Law of the Sea. If this argument is put into effect by China, it would basically render the South China Sea Beijing's private preserve.

28 Yoshiko Sakurai, *Visit of JINF Delegation to Washington, D.C.*, Japan Institute for National Fundamentals, April 2009, available at en.jinf.jp/news/archives/177 (accessed 16 July 2012).

29 Keith Bradsher, "China Reportedly Plans Strict Goals to Save Energy", *The New York Times*, 4 March 2011, available at www.nytimes.com/2011/03/05/business/energy-environment/05energy.html?pagewanted=all (accessed 16 July 2012).

30 "Energy Statistics: The World Gets Back to Burning", Schumpeter Blog, *The Economist* (London) 8 June 2011, available at www.economist.com/blogs/schumpeter/2011/06/energy-statistics (accessed 16 July 2012).

31 Ali Mohammadi, "Iran", *The New York Times*, 3 July 2012, available at topics.nytimes.com/top/news/international/countriesandterritories/iran/index.html (accessed 16 July 2012).

32 On 10 October 1911 an armed uprising led by rebels associated with revolutionary leader Sun on a Qing dynasty garrison. The attack set in motion events that led to the overthrow of imperial rule and raised hopes that China could emerge from a century and a half of national humiliation it had endured at the hands of foreign

powers. The Republic of China was established 75 days later, but its Government fled in disarray to Taiwan in 1949 following the victory of Mao Zedong's communists over Chiang Kai-shek's nationalists in the Chinese civil war. See: "Hu Jintao Urges Unification with Taiwan", *Tapei Times* (Taiwan), 10 October 2011, available at www.taipeitimes.com/News/front/archives/2011/10/10/2003515354 (accessed 16 July 2012).

33 Gillian Wong, "China's Hu Urges Unification with Rival Taiwan", *Associated Press*, 9 October 2011.

34 Madeleine Albright, *Memo to the President Elect* (New York: HarperCollins, 2008).

35 "Reunification by Trade? A Plethora of Free-trade Deals is Driving Taiwan Closer to China", *The Economist*, 6 August 2009, available at www.economist.com/node/14191252 (accessed 16 July 2012).

36 Tom Orlik and Bob Davis, "Beijing Diversifies Away from U.S. Dollar", *Wall Street Journal*, 2 March 2012.

7 From strategies to actions

As noted in previous chapters, over the past two decades the Chinese economy has grown at an annual rate of around 10%, a pace that stood out in 2012, especially as the global economy continued to suffer from the financial meltdown of 2008–09. However, experts note that there are three major problems that could derail China from its impressive track record of consistent economic throughput: 1) a burst in the property bubble; 2) unbalanced rebalancing; and 3) rising political unrest.[1] Even under temporary duress, China's phenomenal hunger for aluminium, cement, copper and steel have in part caused its gas and oil consumption to surge. In fact, in no short measure, this accelerated usage of resources turned it into the "world's second-largest oil importer after the United States".[2] In China, for now, this rapid economic growth and massive consumption of energy and resources growth is being driven by export-oriented manufacturing of low-cost goods to the West. The contribution of domestic consumption and services is comparatively lower. For instance, consumption accounts for 42% of China's gross domestic product (GDP), compared to 68% for the USA, 64% for India, 58% for Europe and 55% for Japan. One could say that Beijing has opted to implement a resource-mobilization model of growth instead of a consumption-led approach, but this is changing. Top officials in Beijing are starting to recognize that overreliance on exporting to high-income countries may come with a series of hidden risks. However, change is difficult, especially for a country guided by one-party rule. Gurcharan Das, former Chief Executive Officer (CEO) of Procter & Gamble in India, notes that "Beijing remains highly suspicious of fast-talking capitalists and entrepreneurs. Also, only about ten percent of credit goes to the private sector in China, even though the private sector employs 40 percent of Chinese workforce".[3]

As the world witnesses the growth of China (and India), a key question emerges: Is there enough oil left in the world to welcome market economies to the elite club of the First World? Resource experts claim that the world has already hit peak oil.[4] As demand meets supply, resulting in higher oil prices, it becomes evident to energy analysts "that the output of conventional petroleum liquids, extracted from crude oil reservoirs, will peak at about 95 million barrels per day in 2012"[5] and then begin an irreversible decline that could constitute a trigger for a bitter competition for the oil that is left underground. The global reserves of crude oil are dwindling as more countries fuel their economic engines.

We will now inspect this more closely. It is a fact that "there are no new vast, untouched reserves of oil sitting close to the earth's surface just waiting to be discovered. Even with phenomenal advances in technology"[6] related to detection and exploration, "no one has made such a profitable discovery in more than 50 years in spite of persistent efforts. International oil companies have scoured the globe in search for oil". The oil that was available right under the surface of the Earth has already been, largely, extracted.

In the interest of clarification, "oil is a non-renewable resource; when a reservoir of oil is depleted, no new oil emerges to take its place"—that is, there is no renewal process. Since approximately 1960, "the rate at which the world has consumed oil has outpaced the rate at which we have discovered new oil fields. Today, we find only about one new barrel of oil for roughly every four barrels that we consume". This is called "discovery deficit". "Meanwhile, it is estimated that the world will consume 120 million barrels of oil a day by 2025, over 50 percent more than consumed in 2001". Therefore, "we, the citizens of the world, are forced to confront a bitter reality in which the world is fast approaching the point at which conventional sources of oil will decline until they are forever gone".[7] This is the rationale that informs China's push towards renewable energy sources. They know they will need critical inputs to move their country forward.[8]

To be sure, as conventional oil is harder to find, so unconventional oil (frontier hydrocarbons) can only be found in small quantities deeply embedded within other components such as mud, rock, tar and sands. Before it can be used in its liquid form, it must first be separated from these components. After separation, it can be processed into by-products. Because easy-to-extract oil is more and more difficult to discover, many oil companies have ventured into unchartered terrains in search of this precious commodity. "China, as well as the rest of the countries in the world, acknowledges that more than 50 percent of the world's remaining conventional oil is found in five countries: Saudi Arabia, Iran, Iraq, Kuwait, and the United Arab Emirates"—none of which are outposts of democracy and freedom. "The rest of the conventional oil is found in comparatively small amounts in 14 other countries":[9] Algeria, Angola, Azerbaijan, Brazil, China, Kazakhstan, Libya, Mexico, Nigeria, Norway, Qatar, Russia, the USA and Venezuela. China is already present in a number of these countries, either directly via investments or indirectly through alliances, equity participations and partnerships.

"Oil production in most of the world has reached, or is nearing, its peak. Production in most Middle Eastern countries, on the other hand, is not expected to peak until 2025. This means that an even greater percentage of the world's remaining oil will be consolidated in just a handful of Middle Eastern nations"[10] and a few other totalitarian regimes.[11] "Outside of the Middle East, the remaining oil", which is obtained unconventionally, is becoming more technologically difficult, expensive and environmentally destructive to acquire. To clarify, unconventional "oil is found offshore below the world's ocean waters, making extraction very risky, environmentally harmful, and destructive to coastal communities. Among the largest unconventional sources of oil are the tar sands of Canada and the shale regions of the Midwestern United States". In addition to the problems listed above, "the process of removing the tar and shale from the earth, extracting the oil, converting it into liquid form, and refining it into gasoline is far more energy-intensive and ozone-depleting than traditional methods of oil production, thus, contributing more to climate change"[12] and ecosystem deterioration.[13]

Notwithstanding, Beijing's strategy to ensure its energy security is to pursue a three-track approach, each of which runs simultaneously: 1) develop and revisit all of the domestic gas and oil fields on the Mainland to extract as much resources (including coal) as possible; 2) move forward with its foreign acquisitions of, and investments in, gas, oil, minerals and natural resources, whether privately by corporate conglomerates or publicly through the use of state-owned gas and oil companies (with the pragmatic use of sovereign funding bodies); and 3) become a world pioneer in the design, development, manufacturing and selling of all of the renewable energy technologies that are so far available (solar, wind, geothermal, etc.).

This last point is important to emphasize, given that:

> China overtook the United States to lead a quarterly index of the most attractive countries for renewable energy projects for the first time in 2010, according to a list compiled by the global accounting firm Ernst & Young. China has set itself a goal of generating 15 percent of its electricity from renewable sources by 2020. It almost doubled consumer subsidies for renewable-power generation in the second half of 2009 to US$545 million. Furthermore, China's rank as the most appealing nation for investing in wind and solar power projects follows the failure of U.S. Congress to pass legislation that would have required utilities to use clean energy [renewable alternatives].[14]

As noted, the "United States is the largest consumer of oil. With just five percent of the world's population, the U.S. uses almost 25 percent of the world's oil every year",[15] with nearly 70% of use for transportation alone.[16] "Americans consume as much oil every year as the combined consumptions of five other countries: China, Germany, India, Japan, and Russia".[17]

Within approximately "10 to 15 years, major oil companies will have depleted their own reserves unless major changes occur".[18] In 2004 the Federal Trade Commission estimated that ExxonMobil and ConocoPhillips would most likely run out of oil in 2017, Chevron in 2016, and Shell and British Petroleum in 2015.[19] Unfortunately for large oil companies, all of the world's remaining oil is more or less spoken for. Governments own the vast majority of what is left in terms of accessible reserves. In addition to the paucity of sites to drill, there is new and growing competition for the oil that is left, from the rising powers of Brazil, Russia, China and India, which are increasing both their consumption and their pursuit of oil abroad.[20]

As the previous six chapters have shown, China is not only reshaping the geopolitical and socio-economic contours of the world owing to its determined and focused quest for resources world-wide, but it is also challenging European countries and the USA on their educational, technological and research leaderships. China has all the benefits of access to plentiful capital, strong government will, and its domestic market is massive and enjoys generous incentives for the adoption of an array of renewable energy options. China expects to retain its dominant position in this and other areas, and much more so if Europe continues to be dragged down by Greece and Spain, and the USA is locked in bi-partisan infighting in the House of Representatives and the Senate.

A web of interconnectedness and interdependence

Because of China's stellar ascendance, the tectonic plates of the international political system will shift for a second time since the end of the Cold War.[21] Europe and the USA are following closely Beijing's assertiveness and new-found confidence in foreign affairs, as well as trying to bring China, other major emerging economies and the developing world under a collaborative framework to address climate change, international terrorism and resource depletion. The latter is particularly critical, given that "developing countries will generate nearly 80 percent of growth in world energy demand by 2020", with the Middle East representing 10% and China representing approximately 30%, according to global consulting firms' forecasts.[22] When it comes to energy resources, the world is tightly interdependent.

China, Europe and North America have a healthy appetite for natural gas. "Natural gas is considered a relatively benign fossil fuel. South-Eastern Peru has several trillion cubic feet of natural gas in its remote jungle. Developing this resource is good for the world's energy security, air quality, and the Peruvian economy". China is fully aware that exploiting its natural gas reserves could "transform Peru from a net importer of energy into a net exporter". Peru is cognizant that this transition can translate into "boosting growth, job creation, and revenues".[23] A shift from oil to gas may lower energy prices. In general terms, when energy costs fall, it brings food prices crashing down as well, because the cost of producing foods is now closely tied to the price of crude oil and natural gas, given that petrochemicals are so widely and heavily utilized in the cultivation of cereals and grains.[24] This is important, owing to rising demand for foods in China. The president of the Earth Policy Institute estimates that if China keeps on growing at a baseline level of 8% per year, by 2030 the per caput income of 1.5 billion Chinese will be the same as the US per caput income in the mid-2000s.[25]

As a whole, China's gas-and-oil-only strategy, put forward into measures, can be partitioned into two spheres: national and international. The national measures include the maximization (as feasible and viable as possible) of domestic production of gas and oil, the expansion of domestic refinery capacity (which comes after production has been secured), expansion of operational and emergency oil storage facilities, policies to reduce the use of oil in power generation (i.e. adoption of alternative forms of energy generation such as solar, wind and geothermal), and make further investment in cutting-edge technologies such as conversion of coals to liquids. International measures are difficult to implement, being bereft of top-down and no-questions-asked approaches of authoritarian regimes. However, international measures consist of a wide diversification of oil import sources (i.e. they reduce reliance on the Middle East and open up more import shares to Africa, Central Asia and Latin America). They also continue to build supportive and long-term infrastructure to facilitate a higher level of imports, development of close economic, diplomatic and political relations with key gas and oil suppliers in order to reduce the probability of supply shocks. Furthermore, they have a more robust financing framework, on behalf of the Chinese Government, to finance (directly and indirectly) state-owned resource exploitation companies.

To secure more solidly national interests and to accomplish the abovementioned measures, Beijing counts on a portfolio of financial tools. These include official development assistance (aid, grants and foreign assistance), export credit loans (lines of credit, loans, insurance and export/import guarantees), and investment finance (equity loans and guarantees, buyout offers and majority acquisitions). Tellingly, China's Ministry of Commerce is in charge of the country's aid activities, not any dedicated bilateral agency for international development.

Actions and initiatives: so much to do, but so little time

This section discusses Beijing's strategies and subsequent actions and initiatives to secure energy sources, raw materials and natural resources around the globe.

Endogenous and exogenous

It is important to highlight that China's endogenous endowments (supply) of energy sources, raw materials and natural resources cannot satisfy its domestic demand. The pace of growth outstrips the supply's ability to feed the manufacturing and industrial engines that drive China's export-led approach to economic progress. Moreover, as the one-party state embraces some aspects of free-market capitalism, China's "domestic firms benefit from spillovers associated with" foreign direct investment "and from the export activities of both domestic and foreign firms".[26] The legitimacy and power of the Chinese Communist Party (CCP) is undergirded by sustained growth, job generation, and improvement of rural and urban lives. Beijing recognizes that energy *insecurity* is a vulnerability that could easily set the country back in the gains so far attained, as well as the ability of competitors to choke energy and resources supplies needed for the growth of the country's economic and political power. This is where exogenous sources come in.

Beijing scouts the globe for locations where there are gas, oil, minerals, timber and rubber, and liaises with resource-rich countries in need of assistance and bilateral agreements so that these can have direct access to China's domestic market. In relying on exogenous sources, China diversifies risk by spreading its dependence among many actors and players, and locks up delivery deals on favourable terms with critical networks of countries that control gas and oil supplies (i.e. the Organization of the Petroleum Exporting Countries—OPEC). Broadly speaking, the combination of endogenous and exogenous sources of energy, materials and resources allows China to design, reassess and implement its evolving national energy strategy as it collects inputs and feedback from economic, political, diplomatic and military domains.

Internal and external

It is now clear that "China offers a distinctive model of economic and political development". The China model represents a coexistence "of a free market approach and an authoritarian state in order to maintain economic growth and political stability".[27] The internal developments in terms of energy and resources have to do with the interaction that takes place among ministries, agencies, offices and bureaux. For instance, the "National Energy Administration (NEA) is responsible for formulating and implementing energy development plans and industrial policies, promoting institutional reform in the energy sector, administering energy sectors including coal, oil, natural gas, power (including nuclear power), new and renewable energy",[28] as well as liaising with the National Energy Commission (NEC) for better coordination in formulating energy strategy and development planning. The NEC determines national energy development strategy, addresses significant issues concerning energy security and energy development, and coordinates major programmes of domestic energy development and global cooperation. It is worth highlighting that "China's energy management transferred from the National Development and Reform Commission (NDRC) to the new National Energy Bureau (NEB)",[29] which is a body more attuned to the complexities of domestic affairs and international relations, as well as being closely supervised by CCP senior officials. Externally, Beijing handles much of its energy and resource coordination via state-owned companies, such as the China Petroleum and Chemical Corporation (Sinopec) and the China National Offshore Oil Corporation (CNOOC). Beijing also receives feedback from field operations undertaken by smaller, private operators, as well as its world-wide embassies, where economic attachés and foreign affairs counsellors work around the clock to skim off information from local business communities to identify trends and prospects.

Aid assistance and economic diplomacy

As Beijing sits on its coffers, filled with foreign exchange reserves (more than $3.3 trillion in March 2012), it enjoys the privilege of offering humanitarian aid and economic assistance to less fortunate "countries in Africa, Asia, Latin America, and the Caribbean". China's fast, efficient, "no strings attached" bilateral approach to aid and assistance is popular in the African region, as is Beijing's preference for infrastructure (by Chinese firms) over governance projects. African officials fear that US or European interference will slow down the assistance process and tie conditions to Chinese aid. In the past, the European Union (EU) angered many African countries when it proposed tri-lateral cooperation. African diplomats and senior government officials note that Africa was better off thanks to China's practical approach to development assistance.[30] As time has passed with classical hegemonic powers, they said that "Africans were frustrated by Western insistence on capacity building, which translated, in their eyes, into conferences and seminars. Africa prefers China's focus on infrastructure and tangible projects"[31] that can be used to boost their economies.

The "notion of 'China's exceptionalism'" in Africa, a prominent feature in Beijing's current continental and bilateral engagement, is understood as a normative modality of engagement that seeks to structure relations such that (although they may remain asymmetrical in economic content) they are, none the less, characterized as equal in terms of recognition of economic gains and political standing (mutual respect and political equality).[32] In terms of economic diplomacy, China uses its growing "trade power to enhance its global influence, especially in Africa, Asia, Latin America and the Caribbean, where it is perceived as an all-weather friend" (in contrast to the USA and the EU, which place conditions on their friendships). "Since opening up its economy in 1978, China's foreign trade has grown by leaps and bounds". It has used its cash wisely in securing a high place by way of soft power. For instance, in 2008 its foreign trade volume exceeded $2.56 trillion. This is "70 times more than what it was a decade ago". This trade capacity is palatable to pliable countries wishing to benefit from China's expanding consumption rates, especially after experiencing "$73 billion annual Sino-African trade, mostly with Angola, Equatorial Guinea, Nigeria, the Republic of Congo and Sudan".

Moreover, "China is Africa's second largest trading partner next only to the U.S. African countries welcome China's low-priced manufactured goods, foreign direct investment as well as development aid for building railroads, dams, schools, roads, hospitals, and fibre-optic networks with no conditionality and no questions on human rights".[33]

State owned and privately owned

As noted previously, "China's state-owned companies, like China itself, are diverse. Many of them would make better partners for American and European multinationals than some of their private-sector counterparts",[34] but because national interests are at play, these alliances are more difficult than expected. For example, China Aluminium Corporation's (ChinalCo) "multibillion-dollar purchase of a stake in Rio Tinto has raised fears about Beijing's agenda for the acquisition of Australia's resources".[35] (Australia has ridden the dragon's tail and has not had a recession since 1991. Even the global financial crisis could not halt Australia's resource-driven economy.) Like ChinalCo, there are 200 large state companies that are primarily in utilities, some in heavy industries and others in resource industries. Whether state-owned companies are efficient or not, they "still dominate the national economy, controlling natural resources, utilities, and many other vital sectors. Their power and influence—particularly their links to the ruling Communist Party", of which there are 80 million members in all of China, or the People's Liberation Army (PLA)—give partners and competitors pause, as well as a certain envy, at the depth and breadth of their operation, with almost no governmental oversight, in terms of environmental affairs and civil rights. We believe that over the next 10 years, as the economy and business climate continue to shift in reaction to world affairs, "the ownership structure of state-owned companies will matter much less than the degree of openness they show in their business" and corporate practices, procedures "and management—that is, their accountability, transparency, and receptiveness to new ideas"[36] and novel challenges.

Effects on asset and oil prices

It is widely believed by energy purchasing agents that China's "go abroad" policy, in terms of energy, materials and resources, has resulted in an unbalanced situation where aggressive Chinese bidding for foreign oil companies has had two effects: higher asset prices and higher oil prices. The failed bid of CNOOC for Unocal drove-up prices for other suitors wishing to acquire the company; also, Sinopec acquired Shell in Angola, while at the same time beating India's largest corporation, Oil and Natural Gas Corporation (ONGC) through the strategic use of a $2 billion soft aid loan in the background. As noted in earlier sections, financing is one of China's comparative advantages, as it not only makes use of cash foreign reserves, but also of low-interest rate financial products to entice sellers to acquiesce. The other tactical move is to make huge loans with the agreement to get repaid in gas and oil. The Chinese Ministry of Commerce calls these deals the "exchanging of oil for infrastructure project" a win-win situation that reduces financial transactions costs and complicated record keeping. Such arrangements can be off the books. As a matter of practice, Beijing and its associated corporations have covered most of Africa with these "sweet-offers" as it has applied it successfully in at least 12 countries, including Algeria, Angola, Chad, Egypt, Gabon, Mauritania, Morocco, Niger, Nigeria, Libya, Sudan and Tunisia. However, not all of these African countries are interested in signing up to the same deals of pumping out oil, as there are monies also to be made as middlemen or in the transport of oil. This is most vividly evidenced by the wide range of activities financed (not only under oil-for-infrastructure ententes), which include classic oil supply contracts, exploration and drilling, development of sites and productions, pipeline construction and maintenance, and oil refineries. What can be learned from this subsector exposure in Africa is that Beijing is not only interested in securing gas and oil supplies to the mainland, but also in becoming as vertically integrated as possible in an effort to mimic the very lucrative business model set out by the Americans and Europeans during their golden years. China, evidently, sees America's decline and a more pluralistic world arising as an opportunity to carve out niches in places willing to be opened up and thirsty for no-strings-attached development options.

While it is true that these accords, agreements, deals, ententes and treaties have grown exponentially since the beginning of the 21st century, these have not occurred without complications and criticism. To begin with, there are the claims of China crowding out the world of oil through its actions and initiatives abroad, but this has turned out to be unfounded, given that aside from higher oil prices, no state has so far suffered from gas and oil shortages owing to China's interventions. Another argument is that Beijing props up dictatorial regimes in Africa and Central Asia, but this is also a flimsy claim, given that dictatorships, demagogy and authoritarianism pervaded Africa and Central Asia prior to China's arrival. Whether this phenomenon has been accentuated or not is a matter of controversy, given that tyrants could as well have taken easy money from Western countries if these were not bound by the studious oversight of civil society and highly regulated democratic institutions. The latter concerns a number of issues raised by international oil conglomerates regarding predatory financing by Chinese financial institutions. Here, the claims are that the Chinese Export-Import Bank has become a covert economic-political instrument through which Beijing and its corporate associates have supported new investments and expansion of private and public oil companies overseas. This, competitors have said, is masked from public scrutiny through direct and indirect subsidies and secretive side deals with high-level state officials in host countries (i.e. in the form of aid programmes linked to gas and oil contracts and other exploitation deals). Competitors go so far as to state that this is a credit race in the making.

Ruminations on progress and vision

Promoted by its reform and open-door policy, China's private sector has grown dramatically over the past 30 years. Furthermore, the private sector displaces the state sector as the dominant player in the economic well-being of China. Some 70% of GDP is now in private hands, even if the private sector employs only 40% of the Chinese workforce. "China's private sector created 11.4 million new jobs in 2009 and contributed to more than 90% of all urban new employments".[37] In total, "the private sector now provides more than 160 million jobs, according to the State Administration for Industry and Commerce".[38] Between 2006 and 2010, "the country registered about 40 million private businesses".[39] "China Mobil and China Telecom are huge, but these are natural monopolies. China is a private-sector economy. In the financial sector, reform has been much delayed. Banking is still a state monopoly, and a concentrated monopoly"[40] at that, but this is set to change in coming years.

In terms of quasi-private sector players, "China has established two of the world's newer large sovereign wealth funds: the official China Investment Corporation (CIC), and the non-official and less transparent State Administration of Foreign Exchange (SAFE) Investment Company [also called SIC]. Both funds provide alternative investment opportunities for China's exploding foreign exchange reserves, at US$3.3 trillion" by March 2012, the largest in world history.[41] More recently, it was reported that CIC is about to be armed with as much as $200 billion in additional cash by the country's treasury. The previous size of the fund was $332.4 billion. Beijing's goals for the fund is to take some of China's foreign reserves and invest them abroad, thus diversifying the country's holdings. Commentators speculate that China is always plotting investments in energy and precious metals, so it is more likely that the fund will be buying gold and oil, rather than shares of Apple or Microsoft, according to commentators.[42]

Now, let us turn our attention to resource usage. In terms of energy utilization for food production, it takes "35 calories of fossil fuel to make one calorie of beef in confinement and 68 calories of fossil fuels to make a calorie of pork".[43] China, with 1.3 billion people, is adopting a richer, meat-based diet that will need to be supplemented from abroad. If this is indeed the case, the intensification of livestock production systems will demand more energy and resources that are already dwindling (and on a lighter note, it is no wonder that people ask, ironically, whether the resulting product will be a food or a fuel). To avoid reaching a carbon dioxide level of 560 parts per million, world leaders will require a massive global energy project aimed at conservation, emission mitigation and oil substitution. This initiative will have to leave some "room for developed countries to grow, using less" fossil fuel, "and for countries like" Brazil, China and India to advance under a progressive pollution cutback programme, "until they fully climb out of poverty and are able to become more energy efficient".[44] These are ambitious ideas that merit consideration given dwindling resources.

China, Europe and the USA should embrace the challenge of developing green economies as decidedly, wholeheartedly and with the same creativity, dedication, entrepreneurship, investment and smarts they once committed in the past to accelerate economic growth at any cost. Comparatively speaking, this is a minor challenge compared to the major setbacks the world confronted and overcame in the last five centuries. In fact, such large-scale projects should proceed as public-private partnerships enjoying strong government and corporate commitments, with engineering, industrial and university leadership across many countries and an internationally binding climate change mitigation framework designed to engage in sustainable and sustained economic growth.[45] Commentators have also pointed to China's evolving water crisis. In other parts of the world this is nothing new. William Finnegan, reporting from Cochabamba, Bolivia, notes that "the world is running out of fresh water, at a time when demand is rapidly increasing ... with protestors fighting the water privatization that has taken place".[46]

Between 1970 and 1990, a small group of countries that produced and exported oil, managed to seize control over the international gas and oil system. These countries managed to restrict supply in order to reap the benefits of much higher prices. The USA, doubtlessly partly motivated by concerns about control of gas and oil reserves in the Middle East, engaged in the Gulf War in 1991, the invasion of Iraq in 2003 and the subsequent occupation of Iraqi territories. Nowadays, the overriding American narrative on China—which is exacerbated by intense media coverage—is that Beijing's discourse of peaceful development in the region deceives "Asian leaders and the international community about China's" true motives, including the replacement of the USA in East Asia. Henceforth, since "state power is based on sustained economic growth" to pay for military strength, CCP officials in Beijing are well aware that no major emerging economy can attain this sustained growth without securing energy supplies and integrating into the globalized capitalist system.[47]

Geoeconomic considerations

After the breakup of the Soviet Union into Russia (plus five newly independent states of the Caspian region)—and after about a decade in making the transition to capitalism—the Russian Federation became a key player in world gas and oil markets. With Moscow, Beijing seeks a stable relationship based on strategic collaboration and pragmatic utilitarianism as both countries depend on gas and oil to sustain economic growth.[48] Also, as China and India maintain ongoing "talks on energy cooperation, the success of these talks will greatly ease competitive frictions between the two Asian titans as their economies grow and expand, and as the demand for gas and oil increases".[49] Additionally, both countries have initiated efforts to resolve tense border claims that led to war in 1962, as well as evolving dialogues on such topics as climate change, water scarcity, terrorism, globalization, radicalism, non-proliferation, and reform of the UN.

Indeed, as China continues to enjoy more than $3.3 trillion in foreign exchange reserves, it will make tactical overtures and "strategic investments through its sovereign wealth funds"[50] and state-owned financial institutions. When one considers China's hunger for gas, oil, coal and other natural resources, these emerge as likely candidates of interest. International scholar Marc Lanteigne notes that "what separates China from other states, and indeed previous global powers, is that not only is it 'growing up' within a milieu of international institutions far more developed than ever before, but more importantly, it is doing so while making active use of these institutions to promote the country's development of global power status".[51] Finally, Beijing's use of soft power in its Asian neighbourhood unfolds without raising hegemonic suspicions.[52]

Although commentators and observers welcome China's economic growth and international assertiveness, a large number of diplomats and government officials feel negatively about the prospect of Beijing significantly increasing its military (army and navy) power. Again, "it is important to underscore that China's dependence on foreign oil—especially from Africa and the Middle East—will make it more concerned and engaged with the sea lanes" used by its gas and oil tankers, such as routes in the Indian Ocean, the Strait of Malacca and the Taiwan Strait. Chinese officials believe that China could face an energy and economic crisis if its gas and oil supply lines were blocked. For some time now, there has been a rumour that China is building up its navy to protect its commercial ships and oil tankers and oversee shared shipping lines. The new naval strategy consists of defensive and offensive approaches. Furthermore, Beijing forges a number of "strategic relationships along the sea lanes from the Middle East to the South China Sea",[53] so that it can protect its commercial and energy interests. For instance, it has closely liaised with Pakistan for infrastructure projects, with Myanmar for establishing radar systems and to build airstrips, and with Bangladesh for naval facilities. Clearly, Beijing wishes to build its own capacity to secure critically important sea lanes, but it also seeks to continue to cooperate with Europe, Indonesia, Japan, Malaysia, Singapore and the USA to keep the straits open. In the end, it seems that CCP officials will have to engage in a range of diplomatic, economic and political measures to ensure a steady supply of energy sources, raw materials and natural resources.[54]

To this end, "China must develop ties that do not flout international standards of good governance and human dignity, or threaten"[55] American or European "security". Under its energy- and resources-based defence and foreign policies, Beijing has become assertive and bold in seeking what it needs in international markets to keep its economy running at full speed. The USA and EU should contemplate the construction of a stable Asian balance of power in the collective interest of China, its neighbours and its competitors. While keeping China's actions in sight, balanced, patient and farsighted leaders are needed to adapt to rapid changes in the global distribution of commercial, economic, political and military powers, without allowing temptations for sabre-rattling to override China's peaceful rhetoric and stated aim of peace and prosperity. Quite frankly, the ascendancy of China is not assured. Beijing faces a host of challenges that could have significant destabilizing effects on regional growth and the global economy if not handled smoothly. For example, one of the most frequent scenarios advanced is that, "if China's growth slows and its unemployment rises, it could face uncontrollable political unrest from both those who lose their jobs and the hundreds of millions of rural peasants still living in the countryside".[56]

Power: the ultimate aphrodisiac

Given its power and prominence, China is resistant to global regulatory schemes, not unlike the USA. Beijing rejects the broadening jurisdiction of the International Criminal Court and, even more so, rejects any commitment to reducing its own emissions of greenhouse gases. In the same vein, the Chinese leadership has been even more averse than the USA to ratifying human rights conventions. At this juncture, China has ratified the Covenant on Civil and Political Rights but insists that it is only bound by its own interpretations of that treaty. Moreover, when China joined the World Trade Organization (WTO) in 2001, it "promised to live up to its obligations as a trading state, but it is strongly opposed to any mixing of trade rules or regulations with environmental or labour standards". Andrew Nathan claims to see "signs that China is gradually easing its insistence on protecting its own sovereignty, but also acknowledges that signs of this are certainly very modest and remain quite equivocal".[57]

Accounts by journalists and foreign commentators note that the authoritarian wall in China is starting to show fissures. This is partly explained by the upstream and downstream effects of the economic slowdown, which, inevitably, affected China given that much of its exports are sent to the USA and to members of the EU. As these countries could not boost consumption by flooding the system with fresh money, China started to slow down its manufacturing and industrial engines which immediately translated into lay-offs, unemployment, insecurity, uncertainty and less confidence in the current governing system consisting of one-party rule. Beijing is adept at "handling" popular revolts in faraway locations, in capturing and imprisoning artists, at managing its economic dealings and in attracting foreign capital, but it remains to be seen if Beijing can stamp out waves of dissatisfaction that may rapidly arise if the Chinese were to encounter economic difficulties. Private-sector strategists and investment bankers are not so sure that China is the place where sustained growth will occur and stable business will flourish, especially because property rights protections and the rule of law remain elusive. Some pundits believe that India's differentiating advantage comes in the form of an "entrenched and vibrant democracy that will ultimately drive India to outperform China socially and economically".[58] In terms of power acquisition and power projection, Beijing will reshape the world, as it has risen to be perceived as a superpower that has the size and the strength to challenge the previously unique hegemony of the USA. In doing so, it assumes the burden of responding to the actions of American authorities and their companies to their overseas ventures, especially to those that clash with the strategic national interests of the USA.

Final remarks

In Washington, those who favour cuts in military spending seem not to understand that the US military exists not simply to win wars, but to present such overwhelming superiority to potential enemies as to prevent having to fight a war in the first place. In short, a mighty US military presence in the Asia-Pacific region may dissuade the Chinese in their hegemonic plans.

As in major cities in Europe and the USA, Beijing dreams of economic growth and social progress—new businesses, well-paying jobs, affordable education and health care, and Chinese leadership in new industries, like biotechnology and clean energy. Ultimately, the CCP will recognize that it must support a stronger emphasis on green technology, which will change the way China produces and consumes energy, and is the strategy most likely to ignite a fast-growing economy while strengthening its natural security and vital strategic interests. China needs to ensure its economic growth, political stability and socio-cultural development in the more competitive, complex, fragmented and fast-changing world of the 21st century. This will require all instruments and tools of diplomacy, as well as careful assessment of strategies and actions: not only as to how these are framed internally, but also as to how these will be interpreted externally.

In parallel, China is advised to continue tinkering with grants, incentives and loans to increase the manufacturing of new clean-energy products and more energy-efficient technologies. This it should do while the Government continues to implement its conventional energy security plans with the incoming inputs from international markets and other countries. Beijing will not stop scavenging the world for energy, materials and resources until it is able to balance its combination of sources, and it will not likely be encouraged to reconsider its ways of doing business until the USA proves that it sees China's superpower rise based on merit and hard work.

As former US President Bill Clinton says, "We live in the most interdependent age in history. People are increasingly likely to be affected by actions beyond their borders, and their borders are increasingly open to both positive and negative crossings: travelers, immigrants, money, goods, services, information, communication, and culture; disease, trafficking of drugs, weapons, and people, and acts of terrorism and violent crime".[59] China, no doubt, takes actions that affect others, and these consequences and outcomes are the ones that are followed and scrutinized by states in the international system. We must all take notice that as long as China assiduously pursues a finite resource that is also critical to old powers, it will have to deal with criticism, finger-pointing, coercion, manipulation and vituperations by fellow states.

It is no mystery to CCP officials in Beijing that "the world's growth pattern is unsustainable, because the way" many countries "produce and use energy and deplete natural" resources causes "climate change", ecological imbalances, "and other environmental problems"[60] (apologies to those who believe that climate change is not anthropomorphic). However, they also know that short of a world-wide collapse, there is no real need to slow down the prosperity wave that is sweeping over the country, especially if other countries are unwilling, first, to convert their growth models to more sustainable ones. Additionally, they take no time to acknowledge that the world is becoming more unstable, as evidenced by the rapid spread of financial crises from country to country and region to region, political upheavals, economic insecurity and a shared vulnerability to radicalism and terrorism. Lastly, senior government decision makers are often quick to affirm that our modern world is too unequal in incomes and wealth, in access to education, health, jobs and markets. While China takes all these realities into account, however, it moves forward with its national priorities: lifting millions of its citizens out of poverty, becoming the locomotive of growth in Asia, enabling businesses and individuals to reach their aspirations and dreams under an alternative growth model, and cherry-picking of foreign affairs with which it wants to deal—not as a global leader (China wants neither to lead nor to follow), but as an active participant that calculates *ex ante* its gains and losses before inserting moral and sentimental considerations. In short, what we are currently witnessing is a powerful China that is conveniently utilitarian and evaluative.

Whether we like it or not, "the world is still organized around nation-states, the decisions that national leaders make and the ones citizens support today determine the possibilities of tomorrow".[61] The CCP in Beijing could very well claim that their formula for prosperity, peace and security is a balanced cocktail of a strong, effective private sector and an assertive, coherent public sector (i.e. government) that work together to promote a vibrant economy which includes job creation, rising incomes, low inflation, increasing exports, and greater energy and resource supplies. For example, the prosperous Chinese cities of Guangzhou and Shanghai have been, for some time, deeply committed to "building networks of cooperation involving the public, private and civil society sectors, and are now creating economic opportunities and charging into the future with confidence".[62]

If there is one thing that readers can take from this book, it might be the following: today, China behaves like any other powerful state in the international system. It tries to fulfil its role as promoter of economic growth, protector of its citizens, defender of its boundaries, overseer of its critical and strategic national interests, and designer of its own destiny. The only difference is that it happens to be located on the other side of the world and it is growing quickly on an alternative model. The fact that growth and stability rely on critical resource inputs only adds impetus to the need to move away from non-renewable into renewable energy—hopefully quickly, and in peace.

Notes

1 Bob Davis, "In Fast-Growing China, Dangers Threaten to Hamper its Success", *The Wall Street Journal*, 11 April 2011.
2 Julio Friedmann and Thomas Homer-Dixon, "Out of the Energy Box", *Foreign Affairs*, Vol. 83, No. 6 (November/December 2004), pp. 72–83, p. 75.
3 Gurcharan Das, "The India Model", *Foreign Affairs*, Vol. 85, No. 4 (2006), pp. 2–16, p. 6.
4 Ugo Bardi, "Peak Oil: The Four Stages of a New Idea", *Energy*, Vol. 34, No. 3 (March 2009), pp. 323–26.

5 See: Moujahed al-Husseini, "World Production of Conventional Petroleum Liquids to 2030: A Comparative Overview", *GeoArabia*, Vol. 14, No. 1 (2009), p. 215–67, www.gulfpetrolink.net/v14n1-Husseini.pdf (accessed 16 July 2012).

6 Antonia Juhasz, "Big Oil's Last Stand", FPIF, *Foreign Policy in Focus*, 22 October 2008, www.fpif.org/articles/big_oils_last_stand (accessed 16 July 2012).

7 Antonia Juhasz and John Donovan, "Big Oil, the Big Survivor", *Asia Times*, 3 November 2008, available at royaldutchshellplc.com/2008/11/03/big-oil-the-big-survivor/ (accessed 16 July 2012).

8 Antonia Juhasz, *The Tyranny of Oil: The World's Most Powerful Industry—And What We Must Do To Stop It* (New York: HarperCollins, 2008).

9 Ibid.

10 Ibid.

11 Kenneth S. Deffeyes, *Hubbert's Peak: The Impending World Oil Shortage* (New Jersey: Princeton University Press, 2001).

12 Antonia Juhasz, "Big Oil, the Big Survivor".

13 Antonia Juhasz, *The Tyranny of Oil.*

14 Alex Morales, "China Beats U.S. on Renewable-Energy Investor Ranking", *Bloomberg*, 8 September 2010, available at www.bloomberg.com/news/2010-09-07/china-supplants-u-s-at-top-of-ernst-young-ranking-for-renewable-energy.html (accessed 16 July 2012).

15 US Census Bureau, US Department of Energy and Energy Information Administration (EIA), "World Oil Balance Chart 2003–7", www.eia.doe.gov.

16 This includes cars, trucks, farm machinery, trains, commercial airplanes, boats, motorcycles and private jets.

17 Energy Information Administration, "Top World Oil Consumers, 2006 Table", www.eia.doe.gov.

18 R.W. Bentley, "Global Oil and Gas Depletion: An Overview", *Energy Policy*, Vol. 30 (2002), pp. 189–205, p. 189.

19 Federal Trade Commission Bureau of Economics, "The Petroleum Industry: Mergers, Structural Change, and Antitrust Enforcement", FTC Staff Study, August 2004, p. 68.

20 Sigfrido Burgos Cáceres and Sophal Ear, "The Geopolitics of China's Global Resources Quest", *Geopolitics*, Vol. 17, No. 1 (2012), pp. 47–79; see also Chapter 2 of this volume.

21 Samuel S. Kim, *China and the World: Chinese Foreign Relations in the Post-Cold War Era* (New York: Perseus Books, 1998). See also: Mark Valencia, "Philippines-China Spat Tests ASEAN Solidarity", *The Japan Times*, 14 May 2012, available at www.japantimes.co.jp/text/eo20120514a5.html (accessed 16 July 2012).

22 McKinsey Global Institute, "Curbing Global Energy Demand Growth: The Energy Productivity Opportunity", McKinsey Global Institute Publication, May 2007, pp. 1–290, p. 116.

23 Sebastian Mallaby, "NGOs: Fighting Poverty, Hurting the Poor", *Foreign Policy* (September/October 2004), pp. 50–58.

24 Michael Klare, "We're On the Brink of Disaster", *Salon.com*, 26 February 2009.

25 Lester R. Brown, *Eco-Economy: Building an Economy for the Earth* (Washington, DC: Earth Policy Institute, 2001).

26 K. Y. Cheung, "Spillover Effects of FDI via Exports on Innovation Performance of China's High-Technology Industries", *Journal of Contemporary China*, Vol. 19, No. 65 (2011), pp. 541–57, p. 541.

27 Suisheng Zhao, "The China Model: Can it Replace the Western Model of Modernization?", *Journal of Contemporary China*, Vol. 19, No. 65 (2011), pp. 419–36.

28 *Main Functions of Departments of the NDRC: National Energy Administration (NEA)*, n.d., available at en.ndrc.gov.cn/mfod/t20081218_252224.htm (accessed 16 July 2012).

29 Ibid.
30 J. Huntsman, "US Embassy Cables: African Countries Prefer Chinese Aid to US-China Cooperation", *The Guardian* (London), 4 December 2010, www.guardian.co.uk/world/us-embassy-cables-documents/248299 (accessed 16 December 2012).
31 Johnathan Glennie, "Can Mali be Weaned Off Aid?", *The Guardian* (London), 17 February 2011, available at www.guardian.co.uk/global-development/poverty-matters/2011/feb/17/mali-end-aid-dependence (accessed 16 July 2012).
32 Chris Alden and Daniel Large, "China's Exceptionalism and the Challenges of Delivering Difference in Africa", *Journal of Contemporary China*, Vol. 20, No. 68 (2011), 21–38.
33 N.D. Batra, "Trade Offers Bundled with Support for Authoritarian Rulers, Arms Sales and Debt Waivers", *Business Standard*, 25 October 2009, available at www.business-standard.com/india/news/nd-batra-chinas-economic-diplomacy/374169/ (accessed 16 July 2012).
34 Jonathan R. Woetzel, "Reassessing China's State-owned Enterprises", *McKinsey Quarterly*, July 2008, available at www.mckinseyquarterly.com/Public_Sector/Management/Reassessing_Chinas_state-owned_enterprises_2149 (accessed 16 July 2012).
35 Ibid.
36 Ibid.
37 "Private Sector Created 11.4 mln New Jobs in 2009", 30 January 2010, available at www.china.org.cn/business/2010-01/30/content_19335086.htm (accessed 16 July 2012).
38 "China's Private Sector Employs 160 million People", 5 February 2011, available at www.china.org.cn/business/2011-02/05/content_21868938.htm (accessed 16 July 2012).
39 Ibid.
40 Online Extra: "China is a Private-Sector Economy", *Bloomberg Businessweek Magazine*, 21 August 2005, available at www.businessweek.com/stories/2005-08-21/online-extra-china-is-a-private-sector-economy (accessed 16 July 2012).
41 Stephen Thomas and Ji Chen, "China's Sovereign Wealth Funds: Origins, Development, and Future Roles", *Journal of Contemporary China*, Vol. 20, No. 70 (2011), pp. 467–78.
42 Gregory White, "China Just Armed its Sovereign Wealth Fund with $200 Billion in Fresh Cash", *Business Insider*, 25 April 2011.
43 Richard Manning, "The Oil We Eat: Following the Food Chain Back to Iraq", *Harper's*, Vol. 308, No. 1845 (February 2004), pp. 39–45.
44 Thomas L. Friedman, "The Power of Green", *The New York Times Magazine*, 15 April 2007.
45 Joel Makower and Cara Pike, *Strategies for the Green Economy: Opportunities and Challenges in the New World of Business* (New York: McGraw-Hill, 2009).
46 William Finnegan, "Leasing the Rain: The World is Running out of Fresh Water, and The Fight to Control it Has Begun", *The New Yorker*, Vol. 78, No. 7 (April 2002), pp. 43–52.
47 G. John Ikenberry, "The Rise of China and the Future of the West", *Foreign Affairs*, Vol. 87, No.1 (2008), pp. 23–37.
48 Bobo Lo, *Axis of Convenience: Moscow, Beijing, and the New Geopolitics* (Washington, DC: Brookings Institution Press, 2008).
49 Fang Zhou, "China, India Forming Strategic Ties", *China Daily (Beijing)*, 18 February 2005; Manjeet Kripalani, Dexter Roberts and Jason Bush, "India and China: Oil Patch Partners?" *Business Week*, 7 February 2005.
50 Mitchell Clark, "Europe's Sovereign Debt Means Zero Growth for the Next Few Years", *Profit Confidential*, n.d., available at www.profitconfidential.com/tag/austerity-measures/page/2/ (accessed 16 July 2012).
51 Marc Lanteigne, *Chinese Foreign Policy: An Introduction* (New York: Routledge, 2009), p. 2.

52 Joshua Kurlantzick, *Charm Offensive: How China's Soft Power is Transforming the World* (New Haven, CT: Yale University Press, 2007).
53 "China Builds Up Strategic Sea Lanes", *The Washington Times*, 17 January 2005, available at www.washingtontimes.com/news/2005/jan/17/20050117-115550-1929r/ (accessed 16 July 2012).
54 "China Military Build-Up Seems U.S.-Focused: Mullen", *Reuters*, 4 May 2009; Joshua Norman, "WikiLeaks: China Hiding Military Buildup, Intentions", *CBS News*, 6 January 2011; Hugh de Santis, "The Dragon and the Tigers: China and Asian Regionalism", *World Policy Journal* (Summer 2005), pp. 23–36.
55 David Zweig and Bi Jianhai, *China's Global Hunt for Energy*, Council on Foreign Relations, September/October 2005, available at www.cctr.ust.hk/materials/working_papers/200509-10_workPaper5.htm (accessed 16 July 2012).
56 "What America Must Do to Compete with China and India", *Business Week*, 22 August 2005.
57 Andrew Nathan, "China and International Human Rights Regime", in Elizabeth Economy and Michael Oksenberg (eds), *China Joins the World* (New York: Council on Foreign Relations, 1999), p. 148.
58 William Nobrega, "Why India Will Beat China", *Business Week*, 22 July 2008.
59 Bill Clinton, *Back to Work: Why We Need Smart Government for a Strong Economy* (New York: Alfred A. Knopf, 2011), p. 12.
60 David Saterlee, "Bill Clinton on Government's Role in Meeting Future Challenges", *Socio Dynamics*, 29 December 2011, available at sociodynamics.org/archives/tag/rsquo (accessed 16 July 2012).
61 Ibid.
62 Ibid.

Bibliography

Adelman, M. A., "The Real Oil Problem", *Regulation*, Vol. 27, No. 1 (2004), pp. 16–21.

Agence France-Presse, "Brazil to be the Fifth Largest Economy by 2016—President Says", Agence France-Presse, 29 December 2010.

Agnew, J., "Emerging China and Critical Geopolitics: Between World Politics and Chinese Particularity", *Eurasian Geography and Economics*, Vol. 51, No. 5 (2010), pp. 569–82.

Ahn, S. H., "Framing Energy Security Between Russia and South Korea? Progress, Problems and Prospects", *Asian Survey*, Vol. 50, No. 3 (2010), pp. 591–92.

Albright, M., *Memo to the President Elect* (New York: HarperCollins, 2008).

Alden, C., "China in Africa", *Survival*, Vol. 47, No. 3 (2005).

Alden, C. and Alves, A.C., "China and Africa's Natural Resources: The Challenges and Implications for Development and Governance", Paper No. 41 (Johannesburg: South African Institute of International Affairs, September 2009).

Alden, C. and Large, D., "China's Exceptionalism and the Challenges of Delivering Difference in Africa", *Journal of Contemporary China*, Vol. 20, No. 68 (2011), pp. 21–38.

Al-Husseini, M., "World Production of Conventional Petroleum Liquids to 2030: A Comparative Overview", *GeoArabia*, Vol. 14, No. 1 (2009), pp. 215–67.

Allena, F., Qianb, J. and Qianb, M., "Law, Finance, and Economic Growth in China", *Journal of Financial Economics*, Vol. 77, No. 1 (2005), pp. 60–64.

Altman, R.C., "The Great Crash, 2008: A Geopolitical Setback for the West", *Foreign Affairs*, Vol. 88, No. 1 (2009), pp. 2–14.

Anderlini, J., "China's Growth Model 'Unsustainable'", *Financial Times*, December 2010.

Andrews-Speed, P., *Energy Policy and Regulation in the People's Republic of China* (The Hague: Kluwer Law International, 2004).

Angop, "China Plays Predominant Role in Angola's Economic Development", *Angop* (Luanda), 3 July 2009.

——"Angolan Ambassador Praises Bilateral Cooperation", *Angop* (Luanda), 4 March 2010.

Areddy, J. T., "Sinopec is Linked to Daimler", *The Wall Street Journal*, 27–28 March 2010.

Ash, L., "China in Africa: Developing Ties", *BBC News*, 4 December 2007.

Babel, M. S. and Wahid, S. M., "Freshwater Under Threat – Southeast Asia", UNEP-AIT Report (Nairobi: United Nations Environment Programme, 2009).

Balint-Kurti, D., "Guinea – Bought by Beijing", *The World Today* (London), March 2001.

Barboza, D., "In Chinese Factories, Lost Fingers and Low Pay", *The New York Times*, 5 January 2008.

——"State-Owned Group Fuel China's Real Estate Boom", *The New York Times*, 1 August 2010a.

——"China Passes Japan as Second-Largest Economy", *The New York Times*, 15 August 2010b.

Bardi, U., "Peak Oil: The Four Stages of a New Idea", *Energy*, Vol. 34, No. 3 (2009), pp. 323–26.

Bast, B., "The Beginning of History", *Newsweek*, 18 April 2011.

Batra, N. D., "Trade Offers Bundled with Support for Authoritarian Rulers, Arms Sales and Debt Waivers", *Business Standard*, 25 October 2009.

Batson, A., "China Seeks More Involvement and More Clout", *The Wall Street Journal*, 31 March 2009.

BBC, "Angola: China's African Foothold", *BBC News*, 20 June 2006.

——"Oil Markets Explained", *BBC News*, 18 October 2007.

——"Brazil Seeks More Access to China to Balance Trade", *BBC News – Business*, 14 February 2012a.

——"Q&A: South China Sea Dispute", *BBC News – Asia Pacific*, 27 June 2012b.

Bentley, R. W., "Global Oil and Gas Depletion: An Overview", *Energy Policy*, Vol. 30 (2002), pp. 189–205.

Bi, J., "China's New Concept for Development, China in a Globalizing World" (Geneva: UNCTAD, 2005), pp. 105–24.

Billon, L. P., "The Political Ecology of War: Natural Resources and Armed Conflicts", *Political Geography*, Vol. 20 (2001), pp. 561–84.

Birmingham, D., *Portugal and Africa* (New York: Palgrave Macmillan, 1999).

Bloomberg, "China is a Private-Sector Economy", *Bloomberg Businessweek*, 21 August 2005.

——"CNOOC Interest in Brazil Assets Driven by China Need for Energy, Chemicals", *Bloomberg News*, 15 September 2010.

Bohr, A., "Central Asia: Responding to the Multi-vectoring Game", in Robin Niblett (ed.), *America and a Changed World: A Question of Leadership* (Hoboken, NJ and London: Wiley-Blackwell and Chatham House, 2010).

Bradley, T., "China's Cambodian Hegemony", *The Diplomat*, 7 May 2009.

Bradsher, K., "China Reportedly Plans Strict Goals to Save Energy", *The New York Times*, 4 March 2011.

Brautigam, D., *The Dragon's Gift: The Real Story of China in Africa* (New York: Oxford University Press, 2009).

Bremmer, I., "Top Risks for 2010" (New York: Eurasia Group, 2012).

Brittain, V., *Death of Dignity: Angola's Civil War* (London: Pluto Press, 1998).

Broedsgaard, K.E. and Heurlin, B., *China's Place in Global Geopolitics: Domestic, Regional and International Challenges* (London: Routledge, 2002).

Brookes, B., "China's Influence in the Western Hemisphere", Heritage Lectures No. 873 (Washington, DC: The Heritage Foundation, 2005).

——"Russia, China, Move in on Latin America" (Washington, DC: The Heritage Foundation, 2008).

Brown, B., "China: Between Global Responsibilities and Internal Transitions", in Robin Niblett (ed.), *America and a Changed World: A Question of Leadership* (Hoboken, NJ and London: Wiley-Blackwell and Chatham House, 2010).

Brown, L. R., *Eco-Economy: Building an Economy for the Earth* (Washington, DC: Earth Policy Institute, 2001).

Brzezinski, Z., "From Hope to Audacity: Appraising Obama's Foreign Policy", *Foreign Affairs*, Vol. 88, No. 1 (2010).

Buckley, B., "China Tells U.S. to Quit as Human Rights Judge", *Reuters*, 10 April 2011.

Bulmer-Thomas, V., "Latin America: Forging Partnerships in a Transformed Region", in Robin Niblett (ed.), *America and a Changed World: A Question of Leadership* (Wiley-Blackwell and Chatham House, 2010).

Burgis, T., "China in $23bn Nigerian Oil Deal", *Financial Times*, 15 May 2010.

Burgos Cáceres, S. and Ear, S., "China's Strategic Interests in Cambodia: Influence and Resources", *Asian Survey*, Vol. 50, No. 3 (2010), pp. 615–39.

——"China's Natural Resource Appetite in Brazil", *Asian Journal of Latin American Studies*, Vol. 24, No. 2 (2011), pp. 71–92.

——"China's Oil Hunger in Angola: History and Perspective", *Journal of Contemporary China*, Vol. 21, No. 74 (2012a), pp. 351–67.

——"The Geopolitics of China's Global Resources Quest", *Geopolitics*, Vol. 17, No. 1 (2012b), pp. 47–79.

Burgos, S. and Otte, M. J., "Poultry, HPAI, and Livelihoods in Cambodia – A Review", FAO of the UN, Mekong Team Working Paper, No. 3 (Rome: FAO, 2008).

BusinessWeek, "What America Must Do to Compete with China and India", *BusinessWeek*, 22 August 2005.

Bustelo, P., "China and the Geopolitics of Oil in the Asian Pacific Region" (Madrid: Real Instituto Elcano de Estudios Internacionales y Estrategicos, 2005).

Callahan, W. A., "National Insecurities: Humiliation, Salvation, and Chinese Nationalism", *Alternatives*, Vol. 29 (2004), pp. 199–218.

Campos, I. and Vines, A., *Angola and China: A Pragmatic Partnership*, 2008, csis.org/files/media/csis/pubs/080603_campos_angolachina.pdf (accessed 11 July 2012), p. 5.

Carmodya, P. and Taylor, I., "Flexigemony and Force in China's Resource Diplomacy in Africa: Sudan and Zambia Compared", *Geopolitics*, Vol. 15, No. 3 (2010), pp. 496–515.

Castañeda, J., "Morning in Latin America", *Foreign Affairs*, Vol. 87, No. 5 (2008).

CBO Financial, "CBO Proves That—Contrary to Popular Belief—It is Easy Being Green", *Algae Technology*, CBO, USA, n.d.

Chakrya, K. S. and Strangio, S., "Villagers Gather to Express Dam Concerns", *Phnom Penh Post*, 24 September 2009.

Chernia, J. A. and Kentishb, J., "Renewable Energy Policy and Electricity Market Reforms in China", *Energy Policy*, Vol. 35, No. 7 (2007), pp. 3616–29.

Chestney, N., "China's CO2 Emissions Rose 10 Percent in 2010: BP Data", *Reuters*, 8 June 2011.

Cheung, K. Y., "Spillover Effects of FDI via Exports on Innovation Performance of China's High-Technology Industries", *Journal of Contemporary China*, Vol. 19, No. 65 (2011), pp. 541–57.

China Auto Web, "Sales information", chinaautoweb.com/category/sales/page/2/ (accessed 16 July 2012).

China Daily, "China-Africa Trade Up 45% in 2008 to US$107 billion", *China Daily* (Beijing), 11 February 2009.

——"China's Electric Consumption Up in July", *China Daily* (Beijing), 14 August 2010.

China Daily, "Clinton Seeks Better Ties with Oil Giant Angola", *China Daily* (Beijing), 10 August 2009.

China.org.cn, "Private Sector Created 11.4 mln New Jobs in 2009", 30 January 2010, www.china.org.cn/business/2010-01/30/content_19335086.htm (accessed 16 July 2012).

——"China's Private Sector Employs 160 million People", 5 February 2011, www.china.org.cn/business/2011-02/05/content_21868938.htm (accessed 16 July 2012).

The China Post, "China's Electricity Consumption up 6%", *The China Post* (Beijing), 7 January 2010.

Christian Science Monitor, "Oil Drilling: ExxonMobil's Big Discovery in Gulf", *Christian Science Monitor*, 8 June 2011.

Clark, C., "Europe's Sovereign Debt Means Zero Growth for the Next Few Years", *Profit Confidential*, 30 May 2012, www.profitconfidential.com/debt-crisis/europes-sovereign-debt-means-zero-growth-for-the-next-few-years/ (accessed 3 September 2012).

Clinton, B., *Back to Work: Why We Need Smart Government for a Strong Economy* (New York: Alfred A. Knopf, 2011).

CNN, "China GDP Grows by 8.7 Percent in 2009", *CNN* (Atlanta), 21 January 2010.

Coatsworth, J. and Williamson, J., "Always Protectionist? Latin American Tariffs from Independence to Great Depression", *Journal of Latin American Studies*, Vol. 36, No. 2 (2004), pp. 205–32.

Cole, B. D., *The Great Wall at Sea: China's Navy Enters the 21st Century* (Annapolis, MD: Naval Institute Press, 2001).

Collier, P., "The Market for Civil War", *Foreign Policy* (May/June 2003), pp. 38–45.

——"Is Aid Oil? An Analysis of Whether Africa Can Absorb More Aid", *World Development*, Vol. 34, No. 9 (2006), pp. 1482–83.

——*The Bottom Billion: Why the Poorest Countries are Failing and What Can be Done About it* (New York: Oxford University Press, 2008).

Collier, P. and Hoeffler, A., "Greed and Grievance in Civil War", *Oxford Economic Papers*, Vol. 56, No. 4 (2004a), pp. 563–95.

——"Violent Conflict", in *Global Problems: Global Solutions*, edited by B. Lomberg (Cambridge: Cambridge University Press, 2004b).

Collier, P. and Dollar, D., *Globalization, Growth and Poverty: Building an Inclusive World Economy* (New York: Oxford University Press, 2002).

Council on Foreign Relations, "The United States and Southeast Asia: A Policy Agenda for the New Administration" (New York: Report of an Independent Task Force, 2001).

Crompton, C. and Wu, Y., "Energy Consumption in China: Past Trends and Future Directions", *Energy Economics*, Vol. 27 (2005), pp. 195–208.

Cui, C. and Shumsky, T., "China's Warning Signs for 'Dr. Copper'", *The Wall Street Journal*, 11 April 2011.

Dalby, S., "Imperialism, Domination, and Culture: The Continued Relevance of Critical Geopolitics", *Geopolitics*, Vol. 13, No. 3 (2008).

Das, G., "The India Model", *Foreign Affairs*, Vol. 85, No. 4 (2006), pp. 2–16.

Davis, B., "In Fast-Growing China, Dangers Threaten to Hamper its Success", *The Wall Street Journal*, 11 April 2011.

Deffeyes, K. S., *Hubbert's Peak: The Impending World Oil Shortage* (Princeton, NJ: Princeton University Press, 2001).

Delamer, G. R., Goldstein, L.J., Malena, J.E. and Porn, G.E., "Chinese Interests in Latin America", in Paul D. Taylor (ed.), *Latin American Security Challenges* (Newport, RI: Naval War College, 2004).

DeSantis, H., "The Dragon and the Tigers: China and Asian Regionalism", *World Policy Journal* (2005), pp. 23–36.
Dhond, A., "Indian Labour Market and Employment: Changing Scenario in Post-Liberalisation Era", *Journal of Research in Arts & Education*, Vol. 1, No. 1 (2012).
Diamond, J., *Guns, Germs and Steel: The Fates of Human Societies* (New York: W.W. Norton, 1999).
——*Collapse: How Societies Choose to Fail and Survive* (London: Penguin, 2006).
Dibb, P., *Toward a New Balance of Power in Asia* (Oxford: Oxford University Press, 1995).
Djamarani, M., "Scramble for Africa", *Petroleum Review* (2008).
Dodson, B., *China Inside Out: 10 Irreversible Trends Reshaping China and its Relationship with the World* (Hoboken, NJ: Wiley-Blackwell, 2010).
Domínguez, J. I., "China's Relations with Latin America: Shared Gains, Asymmetric Hopes", Inter American Dialogue–China, 2006
Dow Jones Newswires, "Brazil CSN CEO Calls for China Mining Investment Restrictions", *Dow Jones Newswires*, 22 August 2010.
Downs, E., "China", Energy Security Series at Brookings Foreign Policy Studies (Washington, DC: Brookings Institution, December 2006).
Downs, E. S., *China's Quest for Energy Security* (Santa Monica, CA: RAND Corporation, 2006).
——"China Security: The Fact and Fiction of Sino-African Energy Relations" (New York: Council on Foreign Relations, 2007).
Duffy, J., *Portuguese Africa* (Cambridge, MA: Harvard Press, 1959).
Dumbaugh, K., "China-U.S. Relations: Current Issues and Implications for U.S. Policy", CRS Report for Congress No. RL33877 (Washington, DC: Congressional Research Service, 2008).
Dumbaugh, K. and Sullivan, M.P., "China's Growing Interest in Latin America", Report 22119 (Washington, DC: Congressional Research Service, 2005).
Ear, S., "Governance and Economic Performance: Credibility, Political Will, and Reform", *Cambodian Economic Review*, Vol. 1, No. 1 (May 2005), pp. 17–49.
——"The Political Economy of Aid and Governance in Cambodia", *Asian Journal of Political Science*, Vol. 15, No. 1 (2007), pp. 68–96.
——"Sowing and Sewing Growth: The Political Economy of Rice and Garments in Cambodia", Stanford Centre for International Development (SCID), Working Paper No. 384 (Stanford, CA: SCID, 2009a).
——"Cambodian 'Justice'", *The Wall Street Journal Asia* (Hong Kong), 1 September 2009b.
Ear, S. and Cáceres, S. B., "Livelihoods and Highly Pathogenic Avian Influenza in Cambodia", *World's Poultry Science Journal*, Vol. 65, No. 4 (2009), pp. 633–40.
Easterly, W., *The Elusive Quest for Growth: Economists' Adventures and Misadventures in the Tropics* (Cambridge, MA: The MIT Press, 2002).
Ebel, R., "China's Energy Future: The Middle Kingdom Seeks its Place in the Sun" (Washington, DC: Centre for Strategic and International Studies, 2005).
Economic Commission for Latin America and the Caribbean, *Preliminary Overview of the Latin American Economies* (Santiago, Chile: ECLAC, 2008).
Economides, M., "China's Oil Imports Continued Upward Climb in 2009", *Energy Tribune*, 26 January 2010.
The Economist, "Special Report: Beyond Doha", *The Economist* (London), Vol. 389, No. 8601 (2008a).

——"China's Resource Hunger", *The Economist* (London), 13 March 2008b.
——"Cambodia's Oil Resources – Blessing or Curse?" *The Economist* (London), 26 February 2009a.
——"Cambodia, Kuwait, and Farmland – Petrodollars v Smallholders", *The Economist* (London), 23 April 2009b.
——"Reunification by Trade? A Plethora of Free-trade Deals is Driving Taiwan Closer to China", *The Economist* (London), 6 August 2009c.
——"Proliferation from North Korea and Iran – Will Russia and China Pitch in?" *The Economist* (London), 10 September 2009d.
——"Rising Angola: Oil, Glorious Oil", *The Economist* (London), 28 January 2010a.
——"The Oil Industry, Beyond the Black Stuff, Big Oil is Being Forced to Rethink its Future", *The Economist* (London), 4 February 2010b.
——"The World this Week", *The Economist* (London), 20–26 March 2010c.
——"Oil Policy in Brazil, Raining on Rio's Parade", *The Economist* (London), 8 April 2010d.
——"China and America: Repelling Borders", *The Economist* (London), 8 April 2010e.
——"The China Model: The Beijing Consensus is to Keep Quiet", *The Economist* (London), 6 May 2010f.
——"The EU-Latin America Summit: Plus ça change", *The Economist* (London), 13 May 2010g.
——"Asian Security Co-operation: Lost Horizon", *The Economist* (London), 10 June 2010h.
——"Rethinking the Third World: Seeing the World Differently", *The Economist* (London), 10 June 2010i.
——"China: Second in Line", *The Economist* (London), 16 August 2010j.
——"Watching China Whizz By", *The Economist* (London), 19 August 2010k.
——"Energy Statistics: The World Gets Back to Burning", Schumpeter Blog, *The Economist* (London), 8 June 2011, www.economist.com/blogs/schumpeter/2011/06/energy-statistics (accessed 16 July 2012).
——"Attitudes Towards Capital Punishment, Who Goes to the Gallows? Concern Over a High-profile Death Penalty", *The Economist* (London), 28 January 2012.
Economy, E., *The River Runs Black: The Environmental Challenges to China's Future* (Ithaca, NY: Cornell University Press, 2004).
——"China's Rise in Southeast Asia: Implications for Japan and the United States", *The Asia-Pacific Journal* (2005).
Economy, E.C. and Lieberthal, K., "Scorched Earth: Will Environmental Risks in China Overwhelm its Opportunities?" *Harvard Business Review* (2007).
Economy, E. and Oksenberg, M., *China Joins the World: Progress and Prospects* (New York: Council on Foreign Relations, 1999).
EEA (European Environment Agency), *Anthropogenic Processes*, n.d., www.eea.europa.eu/themes/water/wise-help-centre/glossary-definitions/anthropogenic-processes (accessed 15 July 2012).
EIA (Energy Information Agency), "Total Petroleum Consumption", n.d., www.eia.gov/cfapps/ipdbproject/IEDIndex3.cfm?tid=5&pid=5&aid=2 (accessed 15 July 2012).
——"Top World Oil Consumers, 2006 Table", 2006, www.eia.doe.gov (accessed 11 July 2010).

EIU (Economist Intelligence Unit), "Manning the Barricades: Who's at Risk as Deepening Economic Distress Foments Social Unrest", Special Report, *The Economist* (London), 2009.

——"Angola: Country Fact Sheet", *The Economist* (London), 16 February 2010.

Energy Compass, "Asian Disruptions", *Energy Compass* (New York), 11 December 2009.

——"Nigeria: Jonathan Struggles for Power", *Energy Compass* (New York), 5 March 2010.

Erikson, D., "A Dragon in the Andes: China, Venezuela, and US Energy Security", *Military Review* (2006), pp. 83–89.

Erikson, D. and Chen, J., "China, Taiwan, and the Battle for Latin America", *The Fletcher Forum of World Affairs*, Vol. 31, No. 2 (2007), pp. 69–89.

Facts and Details, "Chinese Oil and Gas and Natural Gas Imports, Overseas, Sources and Politics, Chinese Oil Imports", April 2012, factsanddetails.com/china.php?itemid=1727&catid=13&subcatid=85 (accessed 11 July 2012).

Fallows, J., "China's Silver Lining", *The Atlantic* (Washington), June 2008.

Farrel, D., Lund, S., Gerlemann, E. and Seeburger, P., "The New Power Brokers: How Oil, Asia, Hedge Funds, and Private Equity are Shaping Global Capital Markets" (San Francisco, CA: McKinsey Global Institute, 2007).

Federal Trade Commission Bureau of Economics, "The Petroleum Industry: Mergers, Structural Change, and Antitrust Enforcement", FTC Staff Study, August 2004.

Finnegan, W., "Leasing the Rain: The World is Running out of Fresh Water, and the Fight to Control it Has Begun", *The New Yorker*, Vol. 78, No.7 (2002), pp. 43–52.

Forero, F., "China's Oil Diplomacy in Latin America", *The New York Times*, 1 March 2005.

Forney, F., "China's Quest for Oil", *Time* (New York), 18 October 2004.

Friedberg, A. L., "Ripe for Rivalry: Prospects for Peace in a Multipolar Asia", *International Security*, Vol. 18, No. 3 (1993/94), pp. 5–33.

Friedman, T., "The First Law of Petropolitics", *Foreign Policy* (2006).

Friedman, T. L., "The Power of Green", *The New York Times Magazine*, 15 April 2007.

——*Hot, Flat and Crowded: Why We Need a Green Revolution and How it Can Renew America* (New York: Farrar, Straus and Giroux, 2008).

Friedmann, S. and Homer-Dixon, T., "Out of the Energy Box", *Foreign Affairs*, Vol. 83, No. 6 (2004), pp. 72–83.

Gienger, V. and Capaccio, T., "China Seeks Military Power in Indian Ocean, Western Pacific, Pentagon Says", *Bloomberg News* (New York), 17 August 2010.

Glennie, J., "Can Mali be Weaned Off Aid?" *The Guardian*, 17 February 2011.

Global Witness, "Country for Sale: How Cambodia's Elite has Captured the Country's Extractive Industries", Investigative Report (London: Global Witness, 2009).

Glosny, M., "Meeting the Development Challenge in the 21st Century: American and Chinese Perspectives on Foreign Aid" (Washington, DC: National Committee on US-China Relations, 2007).

Godwin, P., *China's Defense Modernization: Aspirations and Capabilities* (Alexandria, VA: CNA Corporation, April 2001).

Goldemberg, J., "Ethanol for a Sustainable Energy Future", *Science*, Vol. 315, No. 5813 (2007), pp. 808–10.

Goldenberg, S., "Canada Looks to China to Exploit Oil Sands Rejected by US", *The Guardian*, 14 February 2010.

Goldstein, A., "Parsing China's Rise: International Circumstances and National Attributes", in Robert S. Ross and Zhu Feng (eds), *China's Ascent: Power, Security, and the Future of International Politics* (Ithaca, NY: Cornell University Press, 2008).
Gombrich, E. H., *A Little History of the World* (New Haven, CT: Yale University Press, 2008).
Goodman, P. S., "China Invests Heavily in Sudan's Oil Industry", *The Washington Post*, 23 December 2004.
Gorelick, S. M., *Oil Panic and the Global Crisis: Predictions and Myths* (Hoboken, NJ: Wiley-Blackwell, 2009).
Greenaway, G., *The World Economy: Global Trade Policy 2009* (New Jersey: Wiley-Blackwell, 2010).
Guan, A., "Vietnam-China Relations Since the End of the Cold War", *Asian Survey*, Vol. 38, No. 12 (1998), pp. 1122–23.
Guan, D., Glen, G.P., Weber, C.L. and Hubacek, K., "Journey to World Top Emitter – An Analysis of the Driving Forces of China's Recent CO2 Emissions Surge", *Geophysical Research Letters*, Vol. 36, No. 4 (2009).
Guthrie, D., *Dragon in a Three-Piece Suit: The Emergence of Capitalism in China* (Princeton, NJ: Princeton University Press, 2001).
Haibin, N., "Emerging Global Partnership: Brazil and China", Shanghai Institute for International Studies, 2010, www.siis.org.cn/en/zhuanti_view_en.aspx?id=10099 (accessed 14 July 2012).
Hakim, P., "The Reluctant Partner", *Foreign Affairs*, Vol. 83, No. 1 (2004), pp. 114–23.
Hall, A., "From fome zero to bolsa família: Social Policies and Poverty Alleviation Under Lula", *Journal of Latin American Studies*, Vol. 38, No. 4 (2006), pp. 689–709.
Hall, J. A., "Human Rights and the Garment Industry in Contemporary Cambodia", *Stanford Journal of International Law*, Vol. 36, No. 1 (2000), pp. 119–74.
Halloran, R., "China's Paths to Hegemony", *The Washington Times*, 17 January 2008.
Halper, S., *The Beijing Consensus: How China's Authoritarian Model Will Dominate the Twenty-first Century* (New York: Basic Books, 2010).
Hanson, H., "China, Africa, and Oil", Backgrounder Publication (Washington, DC: Council on Foreign Relations, 2008).
Hauser, C., "Rising Gas and Food Prices Push U.S. Inflation Higher", *The New York Times*, 13 May 2011.
Hazbun, W., *The Middle East through the Lens of Critical Geopolitics: Globalization, Terrorism, and the Iraq War* (2001), hazbun.mwoodward.com/HAZBUN_Lens%20of%20Critical%20Geopolitics.pdf.
Hemerijck, A., "The Institutional Legacy of the Crisis of Global Capitalism", in A. Hemerijck, B. Knapen and E. van Doorne (eds), *Aftershocks: Economic Crisis and Institutional Choice* (Amsterdam: Amsterdam University Press, 2010).
Hensengerth, O., "Money and Security: China's Strategic Interests in the Mekong River Basin", Briefing Paper, ASP BP 2009/02 (London: Chatham House, 2009).
Hirsh, H., "We're No. 11!" *Newsweek*, 12 August 2010.
Hoekman, B. and Prowse, P., "Economic Policy Responses to Preference Erosion: From Trade as Aid to Aid for Trade", World Bank Policy Research, Working Paper No. 3721 (Washington, DC: World Bank, 2005).
Holmes, J. and Yoshihara, T., "China's Naval Build-Up Not Over", *The Diplomat*, 29 September 2010.

HSBC, "Sterling Absolute Return Strategy", *Quarterly Review* (1 October to 31 December 2010), www.assetmanagement.hsbc.com/uk/attachments/investors/qr_sterling.pdf.
Hubbard, G.R. and Navarro, P., *Seeds of Destruction: Why the Path to Economic Ruin Runs Through Washington and How to Reclaim American Prosperity* (New Jersey: Pearson Education Ltd, 2011).
Hughes, C., *The Political Economy of Cambodia's Transition, 1991–2001* (London: Routledge-Curzon, 2003).
——*Dependent Communities: Aid and Politics in Cambodia and East Timor* (Ithaca, NY: Cornell Southeast Asia Program (SEAP), 2009).
Huntsman, J., "US Embassy Cables: African Countries Prefer Chinese Aid to US-China Cooperation", *The Guardian*, 4 December 2010.
Hwang, K., *Face and Favour: Chinese Power Game* (Chicago, IL: University of Chicago Press, 1987).
IEA (International Energy Agency), *Response System for Oil Supply Emergencies* (2010), www.iea.org/textbase/nppdf/free/response_system.pdf (accessed 11 July 2012).
Ikenberry, J. G., "The Rise of China and the Future of the West: Can the Liberal System Survive?", *Foreign Affairs*, Vol. 87, No. 1 (2008), pp. 23–37.
IMF (International Monetary Fund), "Statement at the Conclusion of an IMF Staff Mission to Cambodia", Press Release, No. 09/325, 24 September 2009.
——*China-Africa Trade Figures* (Washington, DC: IMF, March 2010).
International Institute for Environment and Development, FAO and International Fund for Agricultural Development, *Land Grab or Development Opportunity?* (London: Russell Press, 2009).
Jaffe, A. M. and Lewis, S. W., "Beijing's Oil Diplomacy", *Survival*, Vol. 44, No. 1 (2002), pp. 115–34.
Jeffries, I., *Political Developments in Contemporary China: A Guide* (Abingdon: Taylor & Francis, 2008).
Jeldres, J., "China-Cambodia: More Than Just Friends?", *Asia Times Online* (Hong Kong), 16 September 2003.
Jensen, N. and Wantchekon, L., "Resource Wealth and Political Regimes in Africa", *Comparative Political Studies*, Vol. 37, No. 7 (2004), pp. 816–41.
Jiggins, J., "Vulnerabilities and Tensions in Global Food Systems: Increasing Resilience or Fragility?", Paper presented at the UN Food and Agriculture Organization (FAO), Rome, Italy, 3 March 2010.
Johnson, S. E., "China Emerges as World's Fastest Growing Market for Tropical Timber Products", in *Trends in the Tropical Timber Trade* (UNECE/FAO Forest Products Annual Market Review, 2001–02).
Jones, J., "Sinopec to Pay $4.65 billion in Oil Sand Deal", *Reuters*, 12 April 2010.
Juhasz, A., *The Tyranny of Oil: The World's Most Powerful Industry—And What We Must Do To Stop It* (New York: HarperCollins, 2008).
——"Big Oil's Last Stand", *Foreign Policy in Focus*, 2008.
Juhasz, A. and Donovan, J., "Big Oil, the Big Survivor", *Asia Times*, 3 November 2008.
Kagame, P., "Why Africa Welcomes the Chinese", *The Guardian* (London), 2 November 2009.
Kane, T. M. and Serewicz, L.W., "China's Hunger: The Consequences of a Rising Demand for Food and Energy", *Parameters* (2001), pp. 63–75.
Kang, D. C., *China Rising: Peace, Power, and Order in East Asia* (New York: Columbia University Press, 2007).

Khodeli, I., *From Curse to Blessing? Using Natural Resources to Fuel Sustainable Development* (Hoboken, NJ: Wiley-Blackwell, 2009).

Kim, S. S., *China and the World: Chinese Foreign Relations in the Post-Cold War Era* (New York: Perseus Books, 1998).

Klare, M., "End of the Petroleum Age?", *Foreign Policy in Focus*, 26 June 2008.

——"We're On the Brink of Disaster", *Salon.com*, 26 February 2009.

Kleine-Ahlbrandt, S. and Small, A., "China's New Dictatorship Diplomacy: Is Beijing Parting with Pariahs?", *Foreign Affairs* (2008), pp. 1–7.

Kong, B., *China's International Petroleum Policy* (Santa Barbara, CA: Praeger & Co., 2010).

Kosal, K., "Sino-Cambodia Relations", Cambodian Institute for Cooperation and Peace (CICP), Working Paper No. 28 (Phnom Penh: CICP, 2009).

Krauss, C. and Rosenthal, E., "Reliance on Oil Sands Grows Despite Environmental Risks", *The New York Times*, 18 May 2010.

Kreft, H., "China's Quest for Energy", *Hoover Policy Review*, Vol. 139 (2006).

Krijgsman, R., "China Looks to Downstream Deals as Upstream Acquisition Strategy Fails to Deliver", *Evaluate Energy*, 12 April 2012.

Kripalani, M., Roberts, D. and Bush, J., "India and China: Oil Patch Partners?", *BusinessWeek*, 7 February 2005.

Kurlantzick, J., *Charm Offensive: How China's Soft Power is Transforming the World* (New Haven, CT: Yale University Press, 2007).

Lai, H., "China's Oil Diplomacy: Is it a Global Security Threat", *Third World Quarterly*, Vol. 28, No. 3 (2007).

——"China Builds Bridges to Fuel its Engine Room", *Financial Times*, 4 July 2010.

Lanteigne, M., *Chinese Foreign Policy: An Introduction* (New York: Routledge, 2009).

Leggett, K., "China Flexes Economic Muscle Throughout Burgeoning Africa", *The Wall Street Journal*, 29 March 2005.

Lei, W. and Qinyu, S., "Will China Go to War Over Oil?" *Far Eastern Economic Review*, Vol. 169, No. 3 (2006).

Leverett, F. and Noel, P., "The New Axis of Oil", *The National Interest* (Washington, DC and Sacramento, CA: New American Foundation, 2011), in.reuters.com/article/2011/05/24/idINIndia-57228020110524 (accessed 11 July 2012).

Levkowitz, L., McLellan, M. and Warner, J.R., "A Case Study in Chinese Investors' Operations in Angola and Beyond" (Washington, DC: US–China Economic and Security Review Commission, 2009).

Lewis, J., "Climate Change and Security: Examining China's Challenge in a Warming World", *International Affairs*, Vol. 85, No. 6 (2009).

Lewis, L., "China Demands More Say in Global Iron Ore Trade", *The Times* (London), 13 August 2009.

Li, H., "China's Growing Interest in Latin America and its Implications", *Journal of Strategic Studies*, Vol. 30, No. 4–5 (2007).

Li, Y. and Oberheitmann, A., "Challenges of Rapid Economic Growth in China: Reconciling Sustainable Energy Use, Environmental Stewardship, and Social Development", *Energy Policy*, Vol. 37, No. 4 (2009), pp. 1412–22.

Liao, J.X., "A Silk Road for Oil: Sino-Kazakh Energy Diplomacy", *The Brown Journal of World Affairs*, Vol. XII, No. 2 (2006).

Liberal Conspiracy, "Does China Threatens the Environment?", 4 August 2010, liberal conspiracy.org/2010/08/04/does-china-threaten-the-environment/ (accessed 20 September 2010).

Lim, K. F., "On China's Growing Geo-economic Influence and the Evolution of Variegated Capitalism", *Geoforum*, Vol. 41, No. 5 (2010), pp. 677–88.
Liu, G., Yang, C., Hao, T. and Luo, X., *Oil and Gas Resources in China: A Roadmap to 2050* (New York: Springer Publishing Company, 2010).
Liu, J. and Diamond, J., "China's Environment in a Globalizing World", *Nature*, Vol. 435 (2005), pp. 1–2.
Lloyd, T., "Does Aid Create Trade? An Investigation for European Donors and African Recipients", *European Journal of Development Research*, Vol. 12 (2000), pp. 107–23.
Lo, B., *Axis of Convenience: Moscow, Beijing and the New Geopolitics* (Washington, DC: Brookings Institution, 2008).
Luft, G., "Fuelling the Dragon: China's Race into the Oil Market" (Washington, DC: Institute for Analysis of Global Security, 2004), www.iags.org/china (accessed 17 July 2010).
Lum, T., "China's Assistance and Government Sponsored Investments Activities in Africa, Latin America, and Southeast Asia" (Washington, DC: Congressional Research Service, 2009a).
——"Cambodia: Background and U.S. Relations" (Washington, DC: Report, April 2009b).
Luxner, L., "In Caribbean and Latin America, Chinese Dragon Wakes Up", *Seis Continentes* (Summer 2005).
Lyons, J., "Brazil Turns to China to Help Finance Oil Projects", *The Wall Street Journal* (Hong Kong), 18 May 2009.
Macalister, T., "China Puts £60bn into Energy", *The Guardian*, 5 February 2009.
Macauhub, "China's Trade with Portuguese-speaking Countries Doubles in January", *Macauhub* (Macau), 4 March 2010a.
——"Angola Began Year as China's Top Oil Supplier", *Macauhub* (Macau), 8 March 2010b.
McKibben, B., *Deep Economy: Economics as if the World Mattered* (Oxford: Oneworld Publications, 2007).
McKinsey Global Institute, "Curbing Global Energy Demand Growth: The Energy Productivity Opportunity", McKinsey Global Institute Publication (2007), pp. 1–290.
Makower, J. and Pike, C., *Strategies for the Green Economy: Opportunities and Challenges in the New World of Business* (New York: McGraw-Hill, 2009).
Mallaby, S., "NGOs: Fighting Poverty, Hurting the Poor", *Foreign Policy* (2004).
Manning, R., "The Oil We Eat: Following the Food Chain Back to Iraq", *Harper's*, Vol. 308, No. 1845 (2004), pp. 39–45.
Marketos, T. N., *China's Energy Geopolitics: The Shanghai Cooperation Organization and Central Asia* (London: Routledge, 2010).
Marks, P., "China's Cambodia Strategy", *Parameters*, Vol. 30, No. 3 (2000), pp. 92–108.
Mattoo, A. and Subramanian, A., "From Doha to the Next Bretton Woods: A New Multilateral Trade Agenda", *Foreign Affairs*, Vol. 88, No. 1 (2009), pp. 15–26.
Mayner, S., "Chinese Firm Agrees to Buy Brazilian Iron-ore Mine", *The Wall Street Journal*, 27–28 March 2010.
Medeiros, E. S., *China's International Behaviour: Activism, Opportunism and Diversification* (Santa Monica, CA: RAND Corporation, 2009a).
——"Chinese Foreign Policy: The African Dimension", presentation at the FLAD-IPRI conference on "Strategy and Security in Southern Africa" held in Lisbon, Portugal, October 2009b, www.ipri.pt (accessed 16 July 2012).

Medeiros, E. S. and Fravel, M. T., "China's New Diplomacy", *Foreign Affairs* (2003).

Meidan, M., "Shaping China's Energy Policy", *Journal of Contemporary China*, Vol. 18, No. 61 (2009), pp. 603–4.

Meidan, M., Andrews-Speed, P. and Xin, M., "Shaping China's Energy Policy: Actors and Process", *Journal of Contemporary China*, Vol. 18, No. 61 (2009).

Middleton, C., "Cambodia's Hydropower Development and China's Involvement" (Berkeley, CA: International Rivers Report, 2008), pp. 1–77.

Miller, K., "Coping with China's Financial Power", *Foreign Affairs*, Vol. 89, No. 4 (2010).

Ministério dos Petróleos, "Evolução do mercado petrolífero internacional" (Luanda, Angola: Ministério dos Petróleos, 2009), pp. 1–6.

Mitchell, J., *The New Oil Economy: Impacts on Business, Geopolitics and Society* (London: Earthscan, 2001).

——"World Oil Trade–New Oil Axis", *The World Today* (London), March 2010.

Mohammadi, A., "Iran", *The New York Times*, 3 July 2012.

Montgomery, S. L., *The Powers That Be: Global Energy for the Twenty-first Century and Beyond* (Chicago, IL and London: University of Chicago Press, 2010).

Morales, A., "China Beats U.S. on Renewable-Energy Investor Ranking", *Bloomberg*, 8 September 2010.

Motaal, D. A., "Negotiating With Only One China on Climate Change: On Counting Ourselves Lucky", *Yale Journal of International Affairs*, Vol. 6, No.1 (2011), pp. 74–84.

Moxley, M., "Beijing Strengthens Links with Latin America", *Asia Times Online*, 20 August 2010.

MRC (Mekong River Commission), "Meeting the Needs, Keeping the Balance: Strategic Plan 2006–10" (Vientiane: MRC, 2006).

Muller, R. A., *Physics for Future Presidents* (New York and London: W.W. Norton, 2008).

Muskie, E. S., *Exploring Cambodia: Issues and Reality in a Time of Transition* (Centre for National Policy, 1990).

Nanto, D. K. and Manyin, M.E., "China-North Korea Relations" (Washington, DC: Congressional Research Service, 2010).

Nathan, A., "China and International Human Rights Regime", in *China Joins the World*, edited by Elizabeth Economy and Michael Oksenberg (New York: Council on Foreign Relations, 1999).

National Bureau of Statistics of China, "Statistical data", n.d., www.stats.gov.cn (accessed 11 March 2010).

——"Statistical data", n.d., www.stats.gov.cn/english/ (accessed 23 December 2010).

Naughton, B., *The Chinese Economy: Transitions and Growth* (Boston, MA: The MIT Press 2007).

Navarro, P., "Benchmarking the Advantages Foreign Nations Provide their Manufacturers", in *Benchmarking Foreign Advantages* (2009), pp. 135–48.

NDRC (National Development and Reform Commission), *Main Functions of Departments of the NDRC: National Energy Administration (NEA)*, n.d., en.ndrc.gov.cn/mfod/t20081218_252224.htm (accessed 16 July 2012).

Nelson, C., "China's Naval Base Proposal May Raise Suspicion", *The Nation* (Abu Dhabi), 2 January 2010.

Niblett, N., *America and a Changed World: A Question of Leadership* (Hoboken, NJ: Wiley-Blackwell, 2010).

Nicolas, F., "Chinese Direct Investments in Europe: Facts and Fallacies", Briefing Paper (London: Chatham House, 2009).

Nobrega, W., "Why India Will Beat China", *BusinessWeek*, 22 July 2008.

Norman, J., "WikiLeaks: China Hiding Military Buildup, Intentions", *CBS News*, 6 January 2011.

NoticiasFinancieras, "Brazil Destines US$367 Million for Research and Innovation", *NoticiasFinancieras* (Miami), 20 February 2006.

Novo Journal, "Angola and China are Scoring Points in Africa", *Novo Journal* (Luanda), 7 November 2009.

OECD and IDB, "Competition Law and Policy in Brazil: A Peer Review" (Paris: Organisation for Economic Co-operation and Development/Inter-American Development Bank, 2010).

Offshore Technology, "Nigeria's Change of Gear", 12 April 2010, www.offshore-technology.com/features/feature81868/ (accessed 3 September 2012).

OPEC (Organization of the Petroleum Exporting Countries), "Monthly Oil Market Report–February 2010" (Vienna, Austria: OPEC Research Division, 2010).

Orlik, T. and Davis, B., "Beijing Diversifies Away From U.S. Dollar", *The Wall Street Journal*, 2 March 2012.

People's Daily Online, "China, Thirsty for Oil, Looks to Central Asian Neighbours", *People's Daily Online*, 20 April 2010.

Perlez, J., "China Emerges as Major Player in Asian Aid, Asia-Pacific", *The New York Times*, 17 September 2006.

——"Beijing Exhibiting New Assertiveness in South China Sea", *The New York Times*, 31 May 2012.

Polaski, S., "Worker Rights and Global Trade: The U.S.-Cambodia Bilateral Textile Trade Agreement", *Harvard Business School Case Studies* (Cambridge, MA: Harvard Business School, 2004).

——"Combining Global and Local Forces: The Case of Labour Rights in Cambodia", *World Development*, Vol. 34, No. 5 (2006), pp. 919–32.

Pomfret, J., "China Invests Heavily in Brazil, Elsewhere in Pursuit of Political Heft", *The Washington Post*, 26 July 2010.

Portugal News, "Angola exporta 29.9 bilioes de dolares em petroleo", *Portugal News*, 8 August 2007.

Power, P., "Geopolitics and Development: An Introduction", *Geopolitics*, Vol. 15, No. 3 (2010), pp. 433–40.

Power, P. and Mohan, G., "Towards a Critical Geopolitics of China's Engagement with African Development", *Geopolitics*, Vol. 15, No. 3 (2010), pp. 462–95.

Prevent or Protect—Cambodia, *Background of the Conflict*, n.d., www.preventorprotect.org/overview/cambodia.html (accessed 15 July 2012).

Pujol, P., "VW in China, EU-China Business Management Training Project", MBA 2010, www.ceibs.edu/images/bmt/research/2012/02/08/4F25743E1D989EC60C4CE797F833670B.pdf (accessed 16 June 2012).

Pye, L. W., *The Dynamic of Chinese Politics* (Santa Monica, CA: RAND Corporation, 1981).

Razalli, F. M., "China's Foreign Energy Quest and the Changing Security Dynamics in ASEAN", Asian Conflicts Reports, Issue 11 (Alexandria, VA: Council for Asian Transnational Threat Research, 2010).

Reid, M., *Forgotten Continent: The Battle for Latin America's Soul* (New Haven, CT: Yale University Press, 2007).

Reuters, "China Military Build-Up Seems U.S.-Focused: Mullen", *Reuters*, 4 May 2009.
——"Vale Sees Iron-ore Output Up 50% by 2015", *Reuters*, 18 January 2011a.
——"Goldman Cuts China 2011, 2012 GDP Growth Forecasts", *Reuters*, 24 May 2011b.
Richardson, M., "Another 'Pearl' in Beijing's String of Ports", *The Japan Times*, 19 August 2010.
Richardson, S., *China, Cambodia, and the Five Principles of Peaceful Co-existence* (New York: Columbia University Press, 2009).
Rogers, J., *A Bull in China: Investing Profitable in the World's Greatest Market* (New Jersey: Wiley-Blackwell, 2007).
Rohter, L., "China Widens Economic Role in Latin America", *The Washington Post*, 20 November 2004.
Romer, P. M., "Economic Growth", in David R. Henderson (ed.), *The Concise Encyclopaedia of Economics* (Indianapolis, IN: Liberty Fund Inc., 2007).
Romero, S., "Chavez Says China to Lend Venezuela $20 billion", *The New York Times*, 18 April 2010.
Ross, R. R., *Cambodia: A Country Study* (Washington, DC: Government Printing Office for the Library of Congress, 1987).
Ryan, M., "No More Slow Boat To China For US Liquefied Natural Gas", *AOL Energy*, 27 January 2012.
Sachs, J., "Welcome to the Asian Century", *Fortune*, Vol. 149, No.1 (2004), pp. 53–54.
——*The End of Poverty* (New York: Penguin Press, 2005).
——"The Earth Provides Enough to Meet Everyone's Needs", *The National*, 2 March 2011.
Saichs, A., *Governance and Politics of China* (New York and London: Palgrave Macmillan, 2004).
Sakurai, Y., "Visit of JINF Delegation to Washington, D.C.", Japan Institute for National Fundamentals, April 2009, en.jinf.jp/news/archives/177 (accessed 16 July 2012).
Salameh, M. G., "China's Global Oil Diplomacy: Benign or Hostile?" (Cleveland, OH: International Association for Energy Economics, 2010).
Samuels, D., "From Socialism to Social Democracy: Party Organization and the Transformation of the Workers' Party in Brazil", *Comparative Political Studies*, Vol. 37, No. 9 (2004), pp. 999–1024.
Saterlee, D., "Bill Clinton on Government's Role in Meeting Future Challenges", *Socio Dynamics*, 29 December 2011, sociodynamics.org/archives/tag/rsquo (accessed 16 July 2012).
Schneider, A.C., "Seven Flaws in China's Growth Model", *The Kiplinger Letter*, 23 August 2010.
Schueneman, T., "Memo Reveals China Fears Developed Nation 'Conspiracy' at COP15 Climate Conference", 22 February 2010, www.globalwarmingisreal.com/blog/2010/02/22/memo-reveals-china-fears-developed-nation-conspiracy-at-cop15/ (accessed 20 September 2010).
Sembcorp Marine, "1st Jack-up Drilling Rig Order from China's Sinopec International (Hong Kong) Co for PPL Shipyard. Stock Exchange Announcements", 17 September 2008, www.sembcorpmarine.com.sg/index.php?mact=News,cntnt01,detail,0&cntnt01articleid=223&cntnt01category_id=111&cntnt01returnid=78 (accessed 14 July 2012).

Shambaugh, D., "China Engages Asia: Reshaping the Regional Order", *International Security*, Vol. 29, No. 3 (2004/05), pp. 64–99.

——*Power Shift: China and Asia's New Dynamic* (Berkeley, CA: University of California Press, 2005).

Shay, S., "Scientists Announce Trove of Fragile New Species in Mekong", *Time*, 29 September 2009.

Sheikh, H., "Mekong River Dolphins Close to Extinction", *Xinhuanet*, 18 June 2009, news.xinhuanet.com/english/2009-06/18/content_11561164.htm (accessed 15 July 2012).

Shen, B. "New Opportunity for China-ASEAN Trade", *Beijing Review*, Vol. 46 (2003).

Sinopec, "About Sinopec: Our Company", n.d., english.sinopec.com/about_sinopec/ (accessed 3 September 2010).

Skidmore, T.E., "Bi-racial USA vs. Multi-racial Brazil: Is the Contrast Still Valid?", *Journal of Latin American Studies*, Vol. 25, No. 2 (2009), pp. 373–86.

Slater, R., *Seizing Power: The Grab for Global Oil Wealth* (Hoboken, NJ: John Wiley & Sons, 2010).

Smith, J., "World Oil: Market or Mayhem?", *Journal of Economic Perspectives*, Vol. 23, No. 3 (2009), pp. 145–64.

Sogge, D., "Angola Failed Yet Successful", Working Paper 81 (Madrid: FRIDE, 2009).

Sonangol, "Company History", n.d., www.sonangol.co.ao (accessed 11 March 2010).

Steinfeld, E.S., *Forging Reform in China: The Fate of State-owned Industry* (New York: Cambridge University Press, 1998).

Sternberg, J., "The Price of Beijing's Resources Obstinacy, Planners Used to Win Kudos for Farsightedness. But they Keep Missing the Most Important Signal of All", *The Wall Street Journal*, 6 June 2012, online.wsj.com/article/SB100014240527 02303753904577449760270296808.html (accessed 16 July 2012).

Stevens, S., "Prospect for Oil Prices – Up and Down", *The World Today* (London), March 2010.

Stiglitz, J., Sen, A. and Fitoussi, J. P., "Report by the Commission on the Measurement of Economic Performance and Social Progress", Paris-France (2009), www.stiglitz-sen-fitoussi.fr (accessed 11 March 2010).

Stuart, S., *Waste: Uncovering the Global Food Scandal* (New York and London: W.W. Norton, 2009).

Swenson-Wright, J., "East Asia: Searching for Consistency", in Robin Niblett (ed.), *America and a Changed World: A Question of Leadership* (Hoboken, NJ, and London: Wiley-Blackwell and Chatham House, 2010).

Taipei Times, "Hu Jintao Urges Unification with Taiwan", *Taipei Times*, 10 October 2011.

Taylor, B., "Toward Hegemony? Assessing China's Asian Ambitions", *Security Challenges*, Vol. 1, No. 1 (2005), pp. 1–5.

Taylor, I., *China's New Role in Africa* (Boulder, CO: Lynne Rienner, 2009).

——*China and Africa: Engagement and Compromise* (London: Routledge, 2006).

——"China's Oil Diplomacy in Africa", *International Affairs*, Vol. 82, No. 5 (2006), pp. 937–59.

Thomas, S. and Chen, J., "China's Sovereign Wealth Funds: Origins, Development, and Future Roles", *Journal of Contemporary China*, Vol. 20, No. 70 (2011), pp. 467–78.

Thurow, R. and Kilman, S., *Enough: Why the World's Poorest Starve in an Age of Plenty* (New York: Public Affairs, 2009).

The Times, "China World's Biggest Energy User", *The Times* (London), 20 July 2010.
Traub, J., "China's African Adventure", *The New York Times*, 19 November 2006.
Tsui, T. and Tollefson, J., *Language Policy, Culture, and Identity in Asian Contexts* (London: Routledge, 2006).
UN Forum on Forests and Mekong River Commission (MRC), *Mandate*, n.d., www.fao.org/forestry/directory/1856/en/ (accessed 15 July 2012).
UN News Centre, "China and Russia Veto US/UK-backed Security Council Draft Resolution on Myanmar", *UN News Centre*, 12 January 2007.
USBR, "Hydroelectric Power", n.d., www.usbr.gov/power/edu/pamphlet.pdf (accessed 18 April 2010).
US Census Bureau, "Data Finders", n.d., www.census.gov (accessed 18 April 2010).
US Census Bureau, US Department of Energy and Energy Information Administration (EIA), "World Oil Balance Chart 2003–7", 2003–07, www.eia.doe.gov.
US Energy Information Administration, "China: Oil", Country Analysis Briefs, n.d., www.eia.doe.gov/cabs/China/Oil.html (accessed 27 August 2010).
——"Saudi Arabia: Oil Exports and Shipping", Country Analysis Briefs, n.d., www.eia.doe.gov/cabs/Saudi_Arabia/OilExports.html (accessed 27 August 2010).
——"China: Coal", Country Analysis Briefs, n.d., www.eia.doe.gov/cabs/China/Coal.html (accessed 28 August 2010).
——"Short-term Energy Outlook", n.d., www.eia.gov/steo/ (accessed 2 August 2011).
Valencia, M., "Philippines-China Spat Tests ASEAN Solidarity", *The Japan Times*, 14 May 2012.
van Dijk, M. P., *The New Presence of China in Africa* (Amsterdam: Amsterdam University Press, 2010).
Vaughn, B. and Morrison, W., "China-South East Asia Relations: Trends, Issues, and Implications for the United States" (Washington, DC: Congressional Research Service, 2006).
Venn, F., *Oil Diplomacy in the Twentieth Century* (New York and London: Palgrave Macmillan, 1986).
Vines, A., Wong, L., Weimer, M. and Campos, I., *Thirst for African Oil: Asian National Oil Companies in Nigeria and Angola* (London: Chatham House Report, August 2009), www.chathamhouse.org.uk (accessed 14 July 2012).
The Wall Street Journal, "China, Russia Strike $25 billion Oil Pact", *The Wall Street Journal*, 18 February 2009.
——"Planners Used to Win Kudos for Farsightedness. But They Keep Missing the Most Important Signal of All", *The Wall Street Journal*, 6 June 2012.
Walsh, J.R., "China and the New Geopolitics of Central Asia", *Asian Survey*, Vol. 33, No. 3 (1993), pp. 272–84.
Warnock, F. E., "Doubts About Capital Controls" (New York: Council on Foreign Relations Press, 2011).
The Washington Times, "China Builds up Strategic Sea Lanes", *The Washington Times*, 17 January 2005.
——"China's Paths to Hegemony", *The Washington Times*, 17 January 2008.
Watson, C., "Adios Taipei, Hola Beijing: Taiwan's Relations with Latin America", *China Brief*, Vol. 4, No. 11 (2004).
Watts, J., Carrington, D. and Goldenberg, S., "Climate Conspiracy", *The Guardian*, 11 February 2010.
Watts, M., "Resource Curse? Governmentality, Oil and Power in the Niger Delta, Nigeria", *Geopolitics*, Vol. 9, No. 1 (2004).

Webber, M., "Re-emerging China and Consequences for Economic Geography", *Eurasian Geography and Economics*, Vol. 51, No. 5 (2010), pp. 583–99.

The Week, "Is China's Cheap-labour Party Over?" *The Week*, 18 June 2010.

Wei, L. T., *Oil and Gas in China: The New Energy Superpower's Relations with its Region* (Singapore: World Scientific Publishing Co., 2010).

Wei, M. and Wong, J., "China to Push Ahead with 'Go Abroad' Policy", *Reuters*, November 2010.

Wesson, R., *The United States and Brazil: Limits of Influence* (New York and Stanford, CA: Praeger and the Hoover Institution, 1981).

White, G., "China Just Armed its Sovereign Wealth Fund with $200 Billion in Fresh Cash", *Business Insider*, 25 April 2011.

White, W., "Special Reports: African Infrastructure 2006", *Financial Times* (London), 21 November 2006.

Wikipedia, "Mao Zedong", n.d., en.wikipedia.org/wiki/Mao_Zedong (accessed 16 July 2012).

——"Racism in Asia", n.d., en.wikipedia.org/wiki/Racism_in_Asia (accessed 15 July 2012).

——"Deng Xiaoping", n.d., en.wikipedia.org/wiki/Deng_Xiaoping (accessed 16 July 2012).

Williamson, J., "Lula's Brazil", *Foreign Affairs*, Vol. 82, No. 105 (2003), pp. 105–7.

Woetzel, J.R., "Reassessing China's State-owned Enterprises", *McKinsey Quarterly*, July 2008.

Wong, G., "China's Hu Urges Unification with Rival Taiwan", *Associated Press*, 9 October 2011.

Woodard, K., *International Energy Relations of China* (Stanford, CA: Stanford University Press, 1981).

World Bank, "Clear Skies, Blue Water: China's Environment in the New Century" (Washington, DC: The World Bank, 1997).

——"Battling the Forces of Global Recession", East Asia and Pacific Update (Washington, DC: World Bank, April 2009).

WWF (World Wildlife Fund), "Mekong Dolphins on the Brink of Extinction" (Washington, DC: WWF, June 2009).

Xiaoji, Q., "World Bank Sees China GDP up 9.5% in 2010", *China Daily* (Beijing), 17 March 2010.

Xinhua News, "New OPEC Chairman Sees Crude Oil at $80 per Barrel Reasonable", *Xinhua News* (Beijing), 23 December 2009.

——"China's Industrial Value-added Output up 14.3 pct in H1", *Xinhua News* (Beijing), 13 July 2011a.

——"China's Industrial Value-added Output up 14 pct in July", *Xinhua News* (Beijing), 9 August 2011b.

Xuetong, Y., "How Assertive Should a Great Power Be?" *The New York Times*, 31 March 2011.

Yoeng, K., "Khmer Rouge Rule", in *Cambodian Culture & History of Khmer*, 16 December 2010.

Zeihan, P., "China: Crunch Time", *Stratfor Global Intelligence* (Austin), 30 March 2010.

Zhang, R., Li, G., Fan, J., Wu, D. L. and Molina, M. J., "Intensification of Pacific Storm Track Linked to Asian Pollution", *Proceedings of the National Academy of Sciences of the USA*, Vol. 104, No. 13 (2007).

Zhao, S., "Nationalism's Double Edge", *The Wilson Quarterly* (2005).

——"China's Global Search for Energy Security: Cooperation and Competition in Asia–Pacific", *Journal of Contemporary China*, Vol. 17, No. 55 (2008).
——"The China Model: Can it Replace the Western Model of Modernization?", *Journal of Contemporary China*, Vol. 19, No. 65 (2011), pp. 419–36.
Zhihong, W., "Wen Heads 'Super Ministry' for Energy", *China Daily* (Beijing), 28 January 2010.
Zhou, F., "China, India Forming Strategic Ties", *China Daily* (Beijing), 18 February 2005.
Zhu, Z., "Petroleum and Power: China, the Middle East, and the United States", *Yale Journal of International Affairs* (2007).
Zweig, D. and Jianhai, B., "China's Global Hunt for Energy", *Foreign Affairs*, Vol. 84, No. 5 (2005).

Websites

China Investment Corporation, www.china-inv.cn/cicen/
Chinese Ministry of Commerce, english.mofcom.gov.cn
CNOOC Limited, www.cnoocltd.com
OECD (Organisation for Economic Co-operation and Development), www.oecd.org
UNCTAD (United Nations Conference on Trade and Development), www.unctad.org
UNFCC (United Nations Framework Convention on Climate Change), www.unfccc.int
US Department of Energy, www.energy.gov

Index

For Product Safety Concerns and Information please contact our EU representative GPSR@taylorandfrancis.com
Taylor & Francis Verlag GmbH, Kaufingerstraße 24, 80331 München, Germany

www.ingramcontent.com/pod-product-compliance
Ingram Content Group UK Ltd.
Pitfield, Milton Keynes, MK11 3LW, UK
UKHW020139200926
475702UK00005B/824
9781857436860